NEUROBEHAVIORAL ANATOMY

NEUROBEHAVIORAL ANATOMY

Third Edition

CHRISTOPHER M. FILLEY

UNIVERSITY PRESS OF COLORADO

© 2011 by the University Press of Colorado

Published by the University Press of Colorado
5589 Arapahoe Avenue, Suite 206C
Boulder, Colorado 80303

 The University Press of Colorado is a proud member of
the Association of American University Presses.

The University Press of Colorado is a cooperative publishing enterprise supported, in part, by
Adams State College, Colorado State University, Fort Lewis College, Mesa State College, Metropolitan State College of Denver, University of Colorado, University of Northern Colorado, and
Western State College of Colorado.

∞ The paper used in this publication meets the minimum requirements of the American
National Standard for Information Sciences—Permanence of Paper for Printed Library Materials.
ANSI Z39.48-1992

Library of Congress Cataloging-in-Publication Data

Filley, Christopher M., 1951–
  Neurobehavioral anatomy / Christopher M. Filley. — 3rd ed.
     p. ; cm.
  Includes bibliographical references and index.
  ISBN 978-1-60732-098-2 (pbk. : alk. paper) — ISBN 978-1-60732-099-9 (e-book)
  1. Neurobehavioral disorders. 2. Neuroanatomy. 3. Brain—Localization of functions. 4.
Neuropsychiatry. 5. Neuroanatomy. 6. Clinical neuropsychology. I. Title.
  [DNLM: 1. Brain—anatomy & histology. 2. Behavior—physiology. 3. Brain Diseases—
physiopathology. 4. Cognition Disorders. WL 348]
  RC386.F55 2011
  616.8—dc22

                                                                    2011003820

Design by Daniel Pratt

To my father, Giles Franklin Filley,
whose scholarship remains an inspiration.

CONTENTS

PREFACE TO THE
THIRD EDITION

This book began as a series of Neurobehavior Seminars given to Neurology residents, students, and fellows at the University of Colorado School of Medicine in the late 1980s. Those talks were intended to be selective, practical, and brief, allowing listeners to go away with a reasonable summary of a complex field that they could readily apply in a number of academic settings. Noting at the time that reading about behavioral neurology could be intimidating because of the variety of puzzling concepts and terms as well as the inherent difficulty of capturing in writing the subtleties of altered behavior, I purposely limited the work to a concise description of a group of topics most crucial for the field. I am most gratified that many found the 1995 edition of this book useful as an accessible introduction to behavioral neurology, and I have learned with pleasure that the book has helped as a handy reference for clinical questions and board examination preparation. Some even read it simply because it was interesting. The second edition in 2001 was prepared in the same spirit, and a number of relatively minor changes were made that did not expand the book's size.

In this, the third edition of *Neurobehavioral Anatomy*, the reader will immediately note several more notable changes. While still not a big book, this volume is larger than its predecessors, and the content has been substantially modified. Every chapter has been thoroughly revised to reflect new developments, and more depth has been added to elaborate on what was originally a series of expanded lecture notes. I did not wish to restructure the organization, as I believe the original topics are all still worthy of their inclusion in the table of contents, but each chapter has been updated to discuss new knowledge that has appeared in the last several years. These advances include both data from new technologies that permit a closer look at the brain as it mediates behavior and a number of shifts in emphasis and nuance that occur as thinking in the field moves briskly along. It is, in fact, thrilling to witness the rapidly changing and vigorous subspecialty of behavioral neurology working with its collaborative disciplines to reveal with ever greater clarity the rich tapestry of brain-behavior relationships.

The neuroanatomy of behavior as disclosed by the study of brain disorders remains the central focus of this book. Whereas some comments on other areas of neuroscience can be found, the major emphasis is on the structure of the brain—particularly as understood through the concept of distributed neural networks—and how this extraordinary organ can be understood as the source of human behavior and its disturbances. Consistent with this goal, treatment will not be a main theme; without doubt, the treatment of many neurobehavioral disorders has improved substantially, but to keep this book within a reasonable length, that important aspect of behavioral neurology will mostly be deferred to other sources of information.

The book is written from a clinical perspective. I considered including a substantial number of neuroimaging figures to illustrate key points, but opted otherwise because behavioral neurology rests squarely upon the clinical evaluation of altered behavior and not primarily on perusal of structural or functional images of the brain. This is not a neuroradiology book, and the reader can choose from several excellent texts to learn more of this rapidly advancing field. Brain-behavior relationships can be elucidated by careful clinical evaluation and knowledge of neuroanatomy and neuropathology, and reliance on a brain image, while clearly helpful, is no substitute for detailed patient evaluation.

In the preface to the first edition of this book, I referred to the brain as "the most fascinating and impressive biological structure known in the universe." Experience has taught me that physicians and medical scientists in other areas feel just that way about the part of the body they study, and that is of course as it should be. Yet in the years I have done this kind of work, the appeal of the brain as the organ of the mind has not diminished in the slightest, and it is indeed a privilege once again to present a view of how the brain actually makes possible what we regard as our minds. And again, I am most grateful to the patients who serve to inspire our thinking through providing access to their singular human experiences in the process of enduring neurobehavioral dysfunction.

I thank Darrin Pratt at University Press of Colorado for encouraging me to undertake this edition and helping with its gradual yet steady completion. Mel Drisko provided outstanding illustrations that highlight essential details of neuroanatomy. I am also grateful to C. Alan Anderson, David B. Arciniegas, James P. Kelly, Benzi M. Kluger, Bruce H. Price, Jeremy D. Schmahmann, M.-Marsel Mesulam, Kenneth M. Heilman, Bruce L. Miller, Mario F. Mendez, Michael P. Alexander, Antonio R. Damasio, Kirk R. Daffner, Daniel I. Kaufer, Thomas W. McAllister, Kenneth L. Tyler, Steven P. Ringel, Victoria S. Pelak, Mark C. Spitz, John R. Corboy, Laurence J. Robbins, Elizabeth Kozora, C. Munro Cullum, Brian D. Hoyt, Michael Greher, James Grigsby, Steven M. Rao, B. K. Kleinschmidt-DeMasters, Bernardino Ghetti, Michael Wiessberg, Hal S. Wortzel, Kristin M. Brousseau, Jack H. Simon, Mark S. Brown, Jody Tanabe, and Philip J. Boyer, all of whom clarified my thinking across the range of their vast collective expertise.

It must be realized that every behavior
has an anatomy.

NORMAN GESCHWIND

NEUROBEHAVIORAL ANATOMY

CHAPTER ONE

BEHAVIOR AND
THE BRAIN

Human behavior has an enduring appeal. Who among us has not reflected from time to time on how it is that a memory is formed, a sentence produced, or an emotion experienced? What is the origin of the thoughts and feelings that seem so distinctively to characterize the human species? Despite the enormous interest of this subject, however, our knowledge of human behavior is remarkably limited. The principle that the brain is the source of behavior has been acknowledged—with some notable exceptions—since the time of Hippocrates in ancient Greece, but the study of this relatively small organ encased in the skull presents challenges like none other in human biology. Many scientific investigators are deterred by the extraordinary complexity of brain-behavior relationships and, thus, select other areas of inquiry in which meaningful advances—and research grants—are assumed to be more easily attainable. Much of the formal study of behavior is descriptive, and even at this level there are formidable difficulties in the reliable characterization of the observed phenomena. Correlating the vast expanse of human behavior with the intricate neurobiology of the brain in health and disease is still more imposing. This state of relative ignorance is particularly regrettable since a better understanding of behavior could provide limitless benefits both in enhancing the achievements of our species and in reducing its destructiveness. Indeed, a more complete view of behavior as a function of the brain would have important implications for every realm of human activity.

By way of introduction to the core information presented in this book, it will be useful first to consider some philosophical and historical background that influences the study of behavior. Then follows a discussion of selected features of brain anatomy that pertain to neurobehavioral function in general. A brief digression into the intriguing but discredited area of phrenology is then presented as an illustration of the perils of simplistic thinking. Finally, we consider behavioral neurology and its unique viewpoint, hoping to demonstrate how knowledge of brain structure and function is critical to a comprehensive understanding of human behavior.

THE MIND-BRAIN PROBLEM

Traditionally, philosophers have taken a primary role in considering the phenomena of human behavior. The introspective method of thinking about one's own thoughts and feelings was the sole available technique throughout most of human history. Scientific investigation of how and why people act as they do has a rather short history. Only in recent times has there been the development of a systematic empirical approach to the study of behavior, first with the rise of psychology in the nineteenth century (James 1890), and then with the explosive growth of neuroscience in the twentieth (Corsi 1991). These two traditions can be seen as "top down" and "bottom up" to signify their different approaches, and both have made major contributions to our understanding of behavior. Yet it hardly need be stated that these empirical endeavors have not laid to rest ancient philosophical issues. Science has by no means provided answers to all questions about the nature of the mind, and some would maintain that it never can (Horgan 1994). Biology can, however, provide provocative information with which to explore these issues. Although it may seem imprudent for a clinical neuroscientist to indulge in the discussion that follows, there is good reason to suppose that old philosophical problems can be more clearly addressed in the light of new biological knowledge (Young 1987).

One of the oldest and most difficult questions in philosophy is that of the relation of mind to body, commonly known as the mind-body problem. Human beings can reasonably assume that there exists, by virtue of daily experience, a conscious mind and, because of equally evident physical realities, an entity known as the body. Of all body parts, it is also apparent that the brain very likely has the most to do with the mind, and the issue is therefore more precisely called the mind-brain problem. The difficulty arises when one realizes that mental states are clearly subjective, whereas the brain is an objective reality. Consciousness, to most people an obvious, albeit mysterious, human characteristic, does not readily appear to spring from the physical object we recognize as the brain. Many question whether a collection of nerve cells and chemicals can explain the ineffable phenomenon of consciousness, which is often equated with or regarded as akin to such concepts as the soul or spirit. As the philosopher John Searle bluntly poses the mind-brain problem: "How, for example, could this grey and white gook inside my skull be conscious?" (Searle 1984, 15). Consciousness does indeed appear to be the most mystifying feature of the human mind, and establishing it as a property of the brain is by no means straightforward.

Two fundamental solutions have dominated philosophical inquiry into this dilemma. For the sake of simplicity, these may be termed dualism and materialism. Dualism, most notably propounded by René Descartes in the seventeenth century, holds that mind and brain are independent; the famous *Cogito ergo sum* ("I think, therefore I am") asserts the primacy of mind over matter (Descartes 1637) and implies that mental activities are divorced from physical events. Descartes did imagine there to be a point of intersection between the mind and the body and suggested the unpaired pineal gland as the site where the mind receives sensory traffic and acts upon the brain. But his steadfast separation of the immaterial mind from the material brain has exerted enormous influence for hundreds of years.

Materialism, advanced in various ways by thinkers as diverse as John Locke, Bertrand Russell, and Francis Crick, contends in general that mind and body are inseparable; as a result, mental events are nothing more than the expression of the brain's physical activities. Advocates of this "identity theory" argue that the Cartesian division between mental and physical substances is no more than an assertion, in the trenchant phrase of Gilbert Ryle, that there exists a "ghost in the machine" (Ryle 1949). An extreme variant of materialism is B. F. Skinner's behaviorism, an influential movement in twentieth-century American psychology emphasizing the manipulation of behavior by environmental conditions (Skinner 1971), and which, in effect, holds the concept of mind to be irrelevant to the scientific study of behavior.

The mind-brain problem continues to be pursued with vigor. Among modern philosophers who have continued the debate are Karl Popper (Popper and Eccles 1977), an advocate of dualist interactionism, and those who reject dualism, such as Searle (1984, 2004), Patricia Churchland (1986), and Daniel Dennett (1991). In particular, Churchland and Dennett have embraced neuroscience to the extent that they employ the term "mind-brain" to express complete acceptance of the identity of mind and brain (Churchland 1986; Dennett 1991).

At first glance, the dualist position may seem untenable in view of modern conceptions of neuroscience, but difficult problems remain nonetheless.

Prominent among them is the question of free will. Do people act "freely" or under strictly determined laws of physics and chemistry? This dilemma can be more precisely posed as follows: If the mind and brain are in fact identical, and the actions of the brain can eventually be understood and predicted, then where is an escape from the determinist trap into which materialism must fall? Will not all behavior be governed by physical forces, and thus free will be impossible? Here are other questions to which science has not yet offered an answer. Arguments such as these continue to pose for some a significant obstacle to an enthusiastic acceptance of the materialist position.

Notwithstanding the lingering uncertainties raised by dualism, it is difficult to deny the practical utility of the materialist perspective. Advances in science are no less impressive if they pertain to the neural basis of behavior than if they lead to the discovery of penicillin for the treatment of bacterial pneumonia. It is undeniable that investigation of the brain has informed the understanding of a wide range of human behaviors that were previously inexplicable as physical phenomena. In clinical practice, experience with stroke, dementia, or traumatic brain injury patients leaves little doubt that activities of the mind are reliably and often dramatically affected by physical alterations in the brain. The fact that uncertain or inconsistent relationships between brain and behavior continue to challenge neuroscientists—as they clearly do—is testimony to the extraordinary complexity of the brain, not evidence that such relationships do not exist. Although occasional neuroscientists can be found who adopt a dualist position (Penfield 1975; Popper and Eccles 1977), the great majority find that physical events are providing increasingly complete and satisfying explanations for the activities of the mind. As a heuristic principle, the notion that brain events underlie and are directly correlated with mental events has been remarkably productive to date. Without necessarily presuming to answer the thorny philosophical questions introduced above, neuroscience has nevertheless assembled an impressive body of data indicating that the mind's activities are an unequivocal result of the brain's structure and function. In this sense, scientific advances shed light on old problems that, while not solved, at least seem less imposing.

The position taken in these pages derives from an unhesitating embrace of the methods and findings of neuroscience, and therefore follows in the materialist tradition. Although neuroscience cannot comment on a nonphysical reality, there seems little to gain by postulating a spiritual or mystical essence that cannot be reduced to the level of scientific analysis, especially when such complex human capacities as memory, language, and emotion are already yielding to this kind of inquiry. Indeed, as we will see in Chapter 9, a neurology of religion is a plausible approach to understanding a human experience that has traditionally been seen as representing divine influence (Saver and Rabin 1997). In this respect, the dualist tradition does remind us that many mental events have been interpreted as dissociated from any apparent physical basis. Because the task ahead requires developing an understanding of how these mental events are organized by the brain, Searle has recently proposed the idea of "biological naturalism" as a perhaps more harmonious solution to the mind-brain problem (Searle 2004). Whatever the terminology preferred, the proposition that mental events are in fact *caused* by neurobiological processes in the brain has a compelling rationale and much empirical support (Geschwind 1985; Churchland 1986; Dennett 1991; Searle 2004), and there is ample reason to expect that continuing explication of the brain's operations will also unravel the secrets of the mind.

GENERAL FEATURES OF BRAIN ANATOMY

Neuroanatomy has been a foundation of behavioral neurology and continues to provide many insights into the neural organization of human behavior. Just as the elemental motor and sensory functions of the nervous system can be understood as emanating from the operations of brain neurons, so too can the myriad phenomena of cognition and emotion (Mesulam 2000; Kandel, Schwartz, and Jessell 2000). This book is concerned with the anatomy of higher functions, and clinically relevant regions of the brain will be covered in the chapters that follow. As an introduction, however, it will be helpful to begin with some general neuroanatomic features of the brain as they bear upon neurobehavioral concepts; complete accounts of neuroanatomy can be found elsewhere (Nauta and Fiertag 1986; Parent 1996; Nolte 2002).

The human brain is a soft, gelatinous collection of gray and white matter encased in the cranium and weighing about 1,400 grams (roughly three pounds) in the adult. Estimates vary, but there may be 100 billion or more neurons in the brain, and at least ten times this number of glial cells (Kandel, Schwartz, and Jessell 2000). As an indicator of the astonishing degree of connectivity between cerebral neurons, each one makes contact with as many as 10,000 others (Kandel, Schwartz, and Jessell 2000). Interneurons, situated between afferent and efferent neurons, constitute by far the largest class of brain neurons, so that the great majority of the brain's neuronal activity is concerned with the processing and transfer of information that occur between sensory input and motor output (Kandel, Schwartz, and Jessell 2000). In other words, a large quantity of nervous tissue lies interposed between the sensory and motor systems to elaborate the phenomena of behavior.

The brain is made up of the cerebrum, the brainstem, and the cerebellum (Figures 1.1 and 1.2). Most important for the higher functions is the cerebrum, which comprises the paired cerebral hemispheres and the diencephalon, the main components of which are the thalamus and hypothalamus. Why the hemispheres are paired, and why they have distinct functional affiliations in contrast

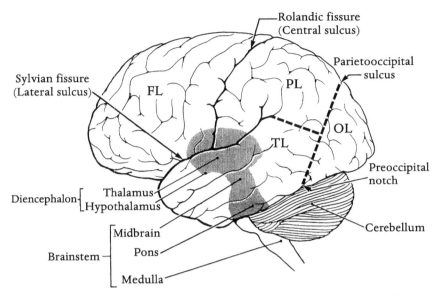

FIGURE 1.1. *Lateral view of the brain depicting lobes and major fissures (FL: frontal lobe; TL: temporal lobe; PL: parietal lobe; OL: occipital lobe).*

to other paired organs in the body such as the lungs and kidneys, are not understood, but the distinct operations of the two cerebral hemispheres will be frequently emphasized in this book. The hemispheres are folded into ridges called gyri, and the grooves between these are known as sulci or fissures. These gross neuroanatomical features form the basis for the division of the hemispheres into four lobes: frontal, temporal, parietal, and occipital.

The parcellation of the hemispheres into four lobes is somewhat arbitrary but serves to produce convenient neuroanatomical landmarks that have important functional affiliations. Table 1.1 gives a brief outline of some prominent brain-behavior relationships, which will be developed in greater detail throughout this book. The frontal lobes, largest and most anterior, provide the origin of the motor system via the corticospinal tracts, mediate the production of language and prosody, and organize the integrative capacities of motivation, comportment, and executive function. The temporal lobes receive primary auditory input, mediate comprehension of language and prosody, and, in concert with the closely connected limbic system, subserve important aspects of memory and emotion. The parietal lobes receive tactile input, mediate visuospatial competence, and subserve reading and calculation skills. The occipital lobes, smallest and most posterior, receive primary visual input and mediate perception of visual material before further processing occurs in more anterior regions.

TABLE **1.1.** Regional functions of the human brain

Frontal Lobes	Parietal Lobes
Motor system	Tactile sensation
Language production (left)	Visuospatial function (right)
Motor prosody (right)	Reading (left)
Comportment	Calculation (left)
Executive function	
Motivation	

Temporal Lobes	Occipital Lobes
Audition	Vision
Language comprehension (left)	Visual perception
Sensory prosody (right)	
Memory	
Emotion	

The hemispheres are connected to each other primarily by the corpus callosum, a massive white matter tract containing some 300 million axons (Nolte 2002; Figure 1.2). This structure permits the continuous interhemispheric exchange of information and joins many distant but homologous cerebral areas into functionally unified networks. The diencephalon is found deep in the brain and has a major role in sensory, motor, arousal, and limbic activities. Within the diencephalon, the egg-shaped thalamus serves as a central relay station for all sensory systems with the exception of olfaction and has a critical role in wakefulness. The tiny hypothalamus exerts enormous influence through its control of the autonomic nervous system, with its sympathetic and parasympathetic divisions, and through its connections with the pituitary gland that enable the neural control of the endocrine system. In posterior and inferior regions of the brain lie the brainstem and the cerebellum. The brainstem, made up of the midbrain, pons, and medulla, plays an essential role in motor and sensory function, and the caudal brainstem contains centers for the control of respiration and cardiac function. The cerebellum acts in combination with gray matter nuclei in the hemispheres and the brainstem known as the basal ganglia (caudate, putamen, globus pallidus, and substantia nigra) to enable fine motor coordination and postural control. At the base of the brain, the medulla exits the skull through the foramen magnum, where it merges with the spinal cord, the most caudal portion of the central nervous system (CNS).

The brain is housed within and protected by the skull, and between the brain and the skull are three membranes: the dura mater, the arachnoid, and the

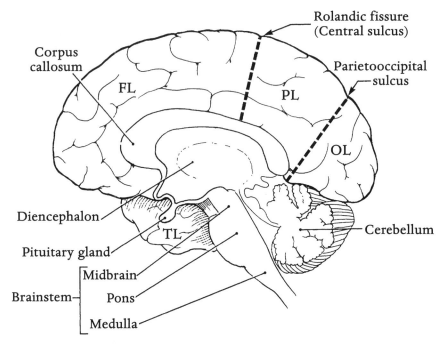

FIGURE **1.2.** *Medial view of the brain depicting the four lobes, diencephalon, brainstem, and cerebellum (FL: frontal lobe; TL: temporal lobe; PL: parietal lobe; OL: occipital lobe).*

pia mater. Within the subarachnoid space, cerebrospinal fluid (CSF) envelops the entire CNS and provides a buoyancy that adds further protection. The CSF is continually produced within the four ventricles of the brain—the paired lateral ventricles in the hemispheres, the third ventricle situated between the two thalami, and the fourth ventricle between the cerebellum and the brainstem—and enters the subarachnoid space through apertures in the fourth ventricle. Eventually the CSF circulates to the vertex of the brain and is absorbed into the venous system through the arachnoid villi. The ventricular system and the CSF are important for the structural support of the brain and for its metabolic activity as well.

The arterial blood supply of the brain originates with two pairs of large vessels in the neck: the internal carotid and the vertebral arteries. The internal carotid arteries then bifurcate into middle and anterior cerebral arteries, which irrigate respectively the lateral hemispheric surfaces and the medial aspects of the frontal and parietal lobes. The vertebral arteries join at the junction of the medulla and the pons to form the basilar artery, which then also bifurcates at the midbrain level to form the two posterior cerebral arteries. These vessels supply the medial and inferior surfaces of the temporal and occipital lobes as well as the caudal diencephalon. Interruption of the blood supply from any of these arteries, as occurs in a stroke, leads to a wide spectrum of important neurobehavioral syndromes. A complex system of cerebral veins conveys blood away from the brain and back to the heart; venous infarction is less common than arterial but can result in similar focal syndromes.

The process of evolution has produced an impressive expansion of the human brain, relative to body weight, in comparison with other animals. There are some species, however—among them some small primates and dolphins—that have proportionately larger brains (Nolte 2002). The size of the brain, therefore, is only one factor accounting for singular human capacities. In humans, the large percentage of the brain devoted exclusively to higher functions is undoubtedly important, as is the exceedingly rich neuronal connectivity of the brain (Nolte 2002). Of all brain regions, the frontal lobes have expanded the most during evolution (Mesulam 2000), and, interestingly, it appears that the main reason for this increase in volume is expansion of frontal white matter (Schoenemann, Sheehan, and Glotzer 2005).

The surface of the brain is called the cortex, from the Latin for "bark," and its regional cytoarchitectonic variations have prompted many attempts to divide it into discrete areas. The most enduring of these cortical maps was devised by the anatomist Korbinian Brodmann (1909). In Figure 1.3, forty-seven cortical areas of Brodmann are depicted, four of which—areas 13 through 16—are not present; these areas, however, actually designate a region called the insula, which is not visible on the outer surface of the brain (Gorman and Unützer 1993). The insula is a small cortical zone buried deep in the Sylvian fissure that is overlain by portions of the frontal, parietal, and temporal lobes known as opercula (*operculum* is Latin for "lid"). Apart from its role in taste perception and some aspects of emotion, the functions of the insula are not well understood. Many of the surface parcellations of Brodmann, however, have well-established functional affiliations, and frequent reference to his schema will be made in this book.

A detailed account of the cerebral cortex is beyond the scope of this book, but selected aspects of cortical structure are relevant. The cortex is a convoluted sheet of gray matter on the outer surface of the brain, much of which is hidden from view in the depths of sulci and fissures. Its thickness ranges between 1.5 and 4.5 mm, with an average of 3 mm. More than 90 percent is made up of neocortex, the phylogenetically recent six-layered cortex that contains about 10 billion of the roughly 100 billion neurons in the brain (Popper and Eccles 1977; Kandel, Schwartz, and Jessell 2000). Other cortical areas, notably the hippocampus and certain olfactory regions linked with the limbic system, have three layers and are known as allocortex.

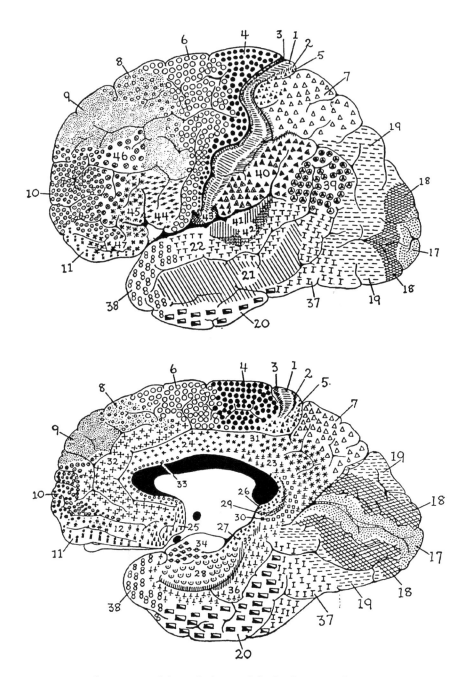

FIGURE 1.3. *Brodmann areas of the cerebral cortex (after Brodmann 1909).*

The neocortex has classically been divided into motor, sensory, and association areas. The last of these is so named because of a philosophical tradition, dating back to the British empiricist philosopher John Locke, holding that mentation is the result of the associations between ideas (Duffy 1984). Association areas occupy the majority of the neocortical surface and, as the presumed repository of ideas, were thus regarded as regions where the higher functions of humans take place, although a certain imprecision about the operations of these regions has always remained.

More recently, neurophysiologic studies of the neocortex have indicated that the fundamental unit of cortical function is a vertically oriented column, perpendicular to the cortical surface, that contains as few as 100 neurons and that is extensively connected with others like it throughout the neocortex (Mountcastle 1978). The most convincing evidence for these columns comes from studies of the primate visual cortex, where it is possible to show with microelectrode insertion that all cells within a column respond similarly to a given external stimulus (Hubel and Wiesel 1977). It is likely, however, that this arrangement is widespread throughout the neocortex (Mountcastle 1978).

These neocortical columns number in the hundreds of millions, and because of the massive connectivity among them, the number of potential combinations between various cortical units is vast indeed. This notion has helped develop the concept of distributed systems in the brain that are composed of large numbers of extensively interlinked modular elements (Mountcastle 1978). The general idea of distributed systems has become so popular that it has largely supplanted the concept of cortical association areas. It should be evident, however, that the two notions are in fact compatible (Duffy 1984), and the shift in terminology reflects advances in basic neuroscience more than a fundamental reconsideration of how the brain functions.

Also important in the elaboration of higher function are numerous structures below the cortical mantle. Gray matter nuclei within the diencephalon, basal ganglia, and brainstem play a special role in fundamental processes such as arousal, attention, and mood. Neurotransmitter systems arise from nuclei deep in the hemispheres or the brainstem and send projections to more rostral sites: the cholinergic system originating from the basal forebrain, the dopaminergic projections from the midbrain substantia nigra and adjacent ventral tegmental area, the noradrenergic system from the pontine locus ceruleus, and the serotonergic fibers from the raphe nuclei of the brainstem. Close structural and functional relationships between subcortical and cortical regions are also readily apparent. Thus, a number of parallel frontal-subcortical circuits have been identified that link cortical and subcortical structures—frontal cortex, basal ganglia, and thalamus—into functional ensembles subserving comportment, motivation, and executive function (Cummings 1993). Figure 1.4 depicts the general organization of these circuits.

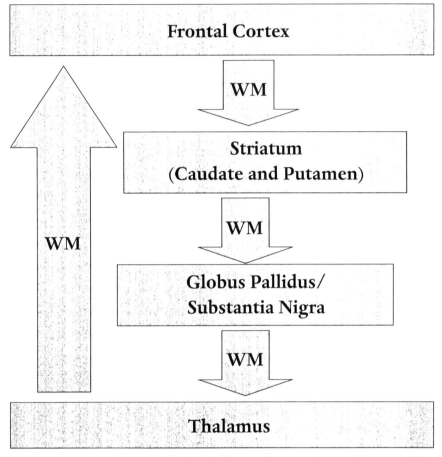

FIGURE 1.4. *General organization of frontal-subcortical circuits (WM: white matter) (after Cummings 1993).*

Not to be neglected are the many white matter tracts that serve to connect cortical and subcortical regions and facilitate rapid and efficient interregional communication (Filley 2001). Intrahemispheric association fibers and interhemispheric commissural fibers link regions within and between the hemispheres, and deeper tracts including the fornix and medial forebrain bundle connect the cerebrum to the limbic system and the brainstem. White matter makes up about one half of the brain volume, and its capacity to facilitate the transfer of information in the brain is a major foundation of normal cognition and emotion (Filley 2001). The collective function of all these subcortical gray and white matter areas thus contributes to the multifocal cerebral ensembles forming the distributed neural networks dedicated to all neurobehavioral domains (Mesulam 1990).

The complex nature of these networks has often invited comparison of the brain to the modern computer, which is increasingly capable of impressive computations that in some ways surpass the abilities of their creators. Science fiction has dealt with some of the implications of computer science, but academicians in the field of artificial intelligence have also been at work, attempting to design machines that mimic if not duplicate the abilities of the human brain (Crevier 1993). Whereas a consideration of computer science and artificial intelligence is beyond the scope of this book, it is instructive to note that the standard computer is known as a *serial* processor, whereas the brain has *parallel* as well as serial processing capacities (Dennett 1991; Mesulam 2000). Both kinds of neural operation are abundantly evident in everyday human life: parallel processing, for example, enables the simultaneous analysis of a continuous flood of sensory and interoceptive input, while serial processing assumes prominence as the most salient stimuli are selected while others are disregarded and an appropriate adaptive response is generated (Mesulam 2000). All of this mental activity features the effortless capacity of the brain to weave disparate aspects of mental experience into a seamless unity, which enables all cognitive and emotional function and culminates in the phenomenon of consciousness. These capacities of the human brain appear to be well beyond the powers of even the most powerful computer. Moreover, given that the phenomenon of consciousness likely depends on billions of living brain cells dynamically connected to each other by trillions of synapses, it is difficult to imagine that a computer will be able to simulate this central feature of the human brain.

Recognition of the imposing problems of artificial intelligence has in fact influenced a turn to more practical uses of the computer, which are in the realm of neuroprosthetics (Friehs et al. 2004). Analogous to the first neural prosthesis introduced for human use—the cochlear implant—a variety of brain-machine and brain-computer interfaces are being explored for clinical application in people with motor disability from such diverse problems as stroke, traumatic brain or spinal cord injury, the locked-in syndrome, multiple sclerosis, amyotrophic lateral sclerosis, muscular dystrophy, and cerebral palsy (Friehs et al. 2004). These laudable efforts promise to assist many people whose problems lie in the transforming of thought into action (Friehs et al. 2004).

THE EXCESSES OF PHRENOLOGY

The scientific study of localizing mental functions in the brain had an inauspicious start. One need only look to the surprisingly popular doctrine in the late eighteenth and early nineteenth centuries known as phrenology, or organology, to appreciate how a too literal approach to brain-behavior relationships can become absurd. Most strongly associated with the names of Franz Joseph Gall (1758–1828) and Johann Kaspar Spurzheim (1776–1832), phrenology claimed to allow the assessment of behavioral traits by simple palpation of the skull, the bumps and ridges thereon allegedly corresponding to anatomical features of the underlying brain that had specific implications such as amativeness, wit, and destructiveness (Gall and Spurzheim 1810–1818). It was thought that the centers for the various traits would develop and expand with regular use, and that this process would actually produce palpable protrusions in the overlying bony surface. Figure 1.5 illustrates the kind of cranial map that resulted from a phrenological interpretation of behavior.

In its heyday, phrenology was widely practiced in Europe and America, and many phrenological societies and journals appeared to advocate the doctrine. Gall, an Austrian physician and neuroanatomist who can legitimately be credited for the recognition that gray matter regions are specifically linked by white matter tracts in the brain, nevertheless assured himself a rather dubious place in medical history by the outlandish assertions of phrenology. Not only was his precise assignment of behavioral characteristics to discrete regions of the brain unsubstantiated, but his belief that cranial prominences could reveal these traits seems remarkably crude and naïve to the modern observer.

Yet the phrenologists should be recognized for mounting formal opposition to the mind-brain dualism that prevailed in their day and beginning to establish a foundation for considering the brain as the organ of behavior. Whereas this insight represented an advance over the orthodoxy of the time, Gall and Spurzheim erred, sadly, in their extremism. The radical localizationism of phrenology was excessive, but the principle of assigning behaviors to regions of the brain was not—as this book will illustrate, a reasonable view of the cerebral localization of behavior is one of the major goals of modern neuroscience, and much has been accomplished to reach this goal.

BEHAVIORAL NEUROLOGY

This book is about the anatomy of human behavior. Use of the word *behavior* will signify the activities of the mind, or, more generally, the total of all the operations ordinarily regarded as mental acts. Neurologists sometimes refer to these as "higher functions," to distinguish them from the elemental sensory and motor functions that are also found in nonhuman species. Some authorities employ the term "higher cortical function" (Luria 1980), although structures in the subcortex such as the thalamus, basal ganglia, and white matter clearly contribute to behavior and thus render this phrase misleading. As a general guideline, these higher *cerebral* functions can be usefully divided into the common-sense catego-

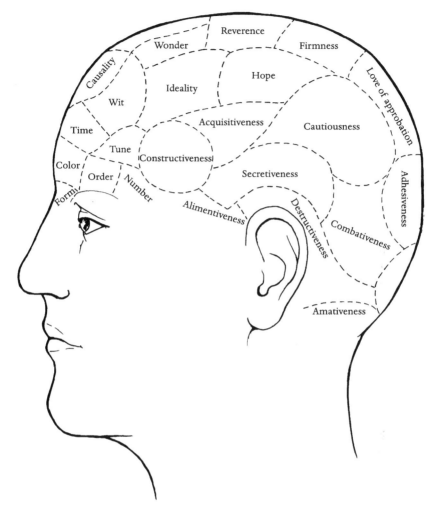

FIGURE 1.5. *Contemporary portrayal of a phrenological map.*

ries of thoughts (cognition) and feelings (emotion), although, in reality, human behaviors usually involve a component of both. Our aim will be to concisely summarize what is known or theorized about the localization of various behavioral capacities in the human brain. This book will concentrate on the adult brain but introduce concepts relevant to the child brain when appropriate.

Our specific approach will rely most heavily on *behavioral neurology* (Damasio 1984), the subspecialty of neurology devoted to the behavioral effects of demonstrable brain disorders, whether from disease, injury, or intoxication. Long an area of interest to neurologists, behavioral neurology has emerged in the last five decades as a subspecialty dedicated to correlating human behavior with the structure and function of the brain (Geschwind 1965; Damasio 1984; Mesulam 2000). This development has been fueled by many factors. First, the hugely influential tradition of Freudian psychoanalysis began to yield around the middle of the twentieth century to a more biological orientation in psychiatry and psychology, and thus opened the door to a more quantitative approach to the study of behavior (Kandel 1979; Price, Adams, and Coyle 2000). Second, advances in neuroscientific technology in the last several decades have kindled the hope that sensitive neuroimaging methods (see below) will help disclose more about the complexities of behavior in health as well as disease by clarifying the structure and function of the brain without the necessity for postmortem examination. Finally, the advent of behavioral neurology as a formal discipline took place largely through the efforts of Norman Geschwind, who in the 1960s established the importance of the field (Geschwind 1965) and influenced a generation of clinician-investigators. Behavioral neurology and related disciplines now stand critically poised at the intersection of modern neuroscience and the detailed clinical observation of behavior. Through study of individuals with focal and diffuse brain lesions, the understanding of contributions made by regions of the brain to various behavioral functions is increasingly complete. Thus the analysis of mind can constructively proceed by way of a clinically based examination of the brain.

A central debate in behavioral neurology has raged between localizationism, the assignment of a given neurobehavioral domain to a specific region of the brain, and holism, the view that cognitive and emotional operations are widely represented in the cerebrum without regional affiliation. We have seen an extreme example of the former in phrenology, whose dubious tradition remains something of an obstacle to a moderate localizationism. In contrast, the holistic viewpoint, also known as the organismic approach or the equipotentiality theory, assumes that the brain works as a unified whole to produce behavior, and that localization is an overly simplistic form of diagram-making. Two syndromes in which holism had vigorous support are aphasia and amnesia, where it was contended that language (Goldstein 1948) and memory (Lashley 1926) disturbances cannot be reliably related to discrete brain areas and that widespread lesions can cause similar defects. The debate is not entirely settled, and it has surely promoted a healthy scientific exchange. However, a consensus has recently been reached that might be called modified localizationism. In this paradigm, widespread interconnected networks of cerebral areas are dedicated to specific functions and not to others (Mesulam 1990, 2000; Cummings 1993); examples include the left perisylvian zone for language and the medial temporolimbic system for memory. These distributed neural networks reflect a conception of brain-behavior relationships that maintains the principle of localizationism without falling prey to its excesses. We shall deal with these networks more completely at several points in this book.

The clinical discipline known as *neuropsychiatry* also deserves comment (Arciniegas and Beresford 2001). Widely popular in nineteenth-century Europe before the ascendancy of Sigmund Freud, neuropsychiatry has recently provoked renewed interest as a discipline having much in common with behavioral neurology (Cummings and Hegarty 1994). Initially, some confusion surrounded this term, however, as the field groped for a secure definition (Caine and Joynt 1986; Yudofsky and Hales 1989; Lishman 1992; Trimble 1993). Since most behavioral neurologists began their training as neurologists and most neuropsychiatrists as psychiatrists, one understandable distinction is that the former group has been more focused on neuroanatomy and neuropathology, whereas the latter has been more concerned with neuropharmacology and psychopathology. Indeed, this book reflects an emphasis on the structural brain correlates of behavior while devoting less attention to neurotransmitter systems and their manipulation by medications. But the two fields are clearly merging. Progress has recently been made by the establishment of the subspecialty of Behavioral Neurology & Neuropsychiatry, which now combines these two traditions into a unified discipline devoted to the evaluation and care of behaviorally disturbed individuals (Arciniegas et al. 2006). This development marks an important milestone in the gradual process of neurology and psychiatry edging closer together in the effort to understand the neurobiological basis of behavior (Kandel 1989; Price, Adams, and Coyle 2000; Cummings and Mega 2003). Indeed, behavioral neurologists and neuropsychiatrists have contributed importantly to clarification of the nosology of many disorders pertinent to both fields in the fourth edition of the American Psychiatric Association's *Diagnostic and Statistical Manual of Mental Disorders* (1994) and its planned successor. These trends are indeed welcome, both for the expanded insights that are likely to appear and for the pressing needs of many patients with disorders of the brain who have not found a comfortable position in either specialty heretofore (Geschwind 1975). Both behavioral neurology and neuropsychiatry have much to contribute to the understanding of behavior, and cooperation is clearly preferable to unproductive interdisciplinary disputes.

Although the clinical study of patients who harbor focal or diffuse brain lesions is the foundation of this book, we will also integrate information derived from the spectacular advances of neuroimaging in recent decades (Naeser and Hayward 1978; Alavi and Hirsch 1991; Posner 1993; Roland 1993; Basser, Mattiello, and LeBihan 1994; Goodkin, Rudick, and Ross 1994; Raichle 1994; Turner 1994; Wright et al. 1995; Prichard and Cummings 1997; Rudkin and Arnold 1999; D'Esposito 2000; Bandettini 2009). It is now possible to study the higher functions in normal and abnormal brains using an array of noninvasive techniques that produce elegant images of brain structure and function. The first advance in structural neuroimaging was *computed tomography* (CT), which appeared in the 1970s and quickly established its role in examining brain-behavior relationships (Naeser and Hayward 1978). Then, in the 1980s, *magnetic resonance imaging* (MRI) offered a more sensitive structural neuroimaging instrument, particularly for viewing the cerebral white matter, and had the additional advantage of avoiding radiation exposure (Goodkin, Rudick, and Ross 1994). More advanced MRI methods were then introduced in the 1990s and are applicable at this point in many areas of research (Bandettini 2009). *Voxel-based morphometry* (VBM; Wright et al. 1995) can be used for measuring gray matter volume; *diffusion tensor imaging* (DTI; Basser, Mattiello, and LeBihan 1994) is helpful in characterizing the structure and location of white matter tracts; and *magnetic resonance spectroscopy* (MRS; Rudkin and Arnold 1999) can be used to measure neurometabolite concentrations in either tissue. For behavioral neurologists, the structural images provided by CT and especially MRI remain most useful and are key components of clinical evaluation (Figures 1.6 and 1.7).

To complement these structural approaches, a variety of methods designed to image the physiology of the brain have been developed, collectively known as functional neuroimaging. *Single photon emission computed tomography* (SPECT; Alavi and Hirsch 1991) and *positron emission tomography* (PET; Roland 1993) exploit the principle that metabolic activity of the brain is coupled with blood flow, so that cerebral function during a cognitive task can be assessed by the introduction of an isotope emitting radiation into the bloodstream. These techniques are useful for identifying areas of the cerebral cortex that become selectively active when the task requires energy expenditure in the responsible neurons. Most recently, *functional MRI* (fMRI) has appeared. Based on the measurement of changes in deoxyhemoglobin concentration that arise from an increase in blood oxygenation in the vicinity of neuronal firing, fMRI has enabled another means of correlating mental operations with brain activity (Prichard and Cummings 1997). These techniques have all proven valuable, and in particular, PET and fMRI have made possible the extraordinary prospect of "seeing the mind" in action (Posner 1993; Turner 1994; Bandettini 2009). fMRI has now become the method of choice in the neuroscience community for functional neuroimaging, as it has both high spatial and temporal resolution and relatively straightforward implementation (Bandettini 2009). As a general rule, functional neuroimaging studies have largely confirmed the brain-behavior relationships discovered from the lesion method upon which behavioral neurology is founded, although methodological problems have hindered progress to some extent. The use of fMRI has enabled many insights about what cortical areas are active in association with a selected cognitive task, but explanations of the mechanisms of cognition

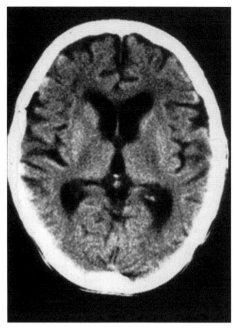

FIGURE 1.6. *Axial computed tomography (CT) scan of a patient with Alzheimer's disease demonstrating mild diffuse cerebral atrophy.*

and how the brain is organized are far more challenging goals that have yet to be accomplished (Bandettini 2009). Nevertheless, functional imaging has made substantial progress, and the use of these increasingly sophisticated instruments doubtless augurs well for continued progress.

Relevant information from related disciplines, including neuropsychology, cognitive science, and the basic neurosciences, will also be integrated into this account. Neuropsychology is a clinical and research discipline with major importance for the precise measurement and characterization of behavior (Lezak et al. 2004). Cognitive science promises many new insights, particularly through its emphasis on computers and theoretical models of cognition (Gardner 1987). Basic neuroscience, despite the lack of an animal model (almost by definition) for higher functions and their disorders, offers valuable information on such matters as neurotransmission and synaptic function (Kandel, Schwartz, and Jessell 2000). An air of excitement pervades the exploration of behavior as interdisciplinary work produces new advances on a daily basis. Despite the many areas of uncertainty in all areas of neuroscience, a growing body of data is providing a cogent description of the roles played by brain structures in the wondrous array of human behaviors.

The study of where and how the higher functions are represented in the human brain is clearly a formidable task. Behavioral neurology can only proceed

FIGURE 1.7. *Axial magnetic resonance imaging (MRI) scan of a patient with Binswanger's disease showing cerebral white matter ischemia.*

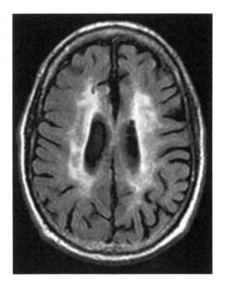

by analyzing abnormal behavior in the context of brain pathology and then theorizing about the normal behavior that has been disturbed (Benson 1994). Difficulties arise both in the definition of neurobehavioral functions, such as attention, memory, language, and the like, and in the precise identification of damaged brain areas, so that both sides of the brain-behavior dyad are subject to significant uncertainties. The lesion method should be used to explore the distributed networks of the brain subserving specific functions and not in an attempt to establish a strict correspondence between the location of a lesion and the site of a mental operation (Damasio and Damasio 1989). The brain is not a collection of lobes, gyri, nuclei, and tracts that each have a single and invariant functional affiliation, and the modern understanding of distributed neural networks teaches that the wide range of cognitive and emotional functions corresponds to an impressively complex neural foundation (Mesulam 2000). With this caveat in mind, it is hoped that a general portrayal of the neuroanatomy of behavior will prove useful. The study of human beings is ultimately the only way to probe their unique behaviors, even though reliance on "nature's experiments"—due to stroke, trauma, degenerative disease, and the like—restricts the scope of such investigation more than tightly controlled experiments with laboratory animals. Our goal will be to examine the domains of mental activity that have been illuminated by clinical observation, demonstrating that such an endeavor not only has great value for the care of individuals with devastating afflictions of the brain but also contributes enormously to the understanding of how human behavior is anatomically organized.

REFERENCES

Alavi, A., and Hirsch, L. J. Studies of central nervous system disorders with single photon emission computed tomography and positron emission tomography: evolution over the past 2 decades. *Semin Nucl Med* 1991; 21: 58–81.

American Psychiatric Association. *Diagnostic and Statistical Manual of Mental Disorders 4th ed.* Washington, DC: American Psychiatric Association; 1994.

Arciniegas, D. B., and Beresford, T. P. *Neuropsychiatry: An Introductory Approach.* Cambridge: Cambridge University Press; 2001.

Arciniegas D. B., Kaufer, D. I., and Joint Advisory Committee on Subspecialty Certification of the American Neuropsychiatric Association and the Society for Behavioral and Cognitive Neurology. Core curriculum for training in behavioral neurology and neuropsychiatry. *J Neuropsychiatry Clin Neurosci* 2006; 18: 6–13.

Bandettini, P. A. What's new in neuroimaging methods? *Ann NY Acad Sci* 2009; 1156: 260–293.

Basser, P. J., Mattiello, J., and LeBihan, D. MR diffusion tensor spectroscopy and imaging. *Biophys J* 1994; 66: 259–267.

Benson, D. F. *The Neurology of Thinking.* New York: Oxford University Press; 1994.

Brodmann, K. *Vergleichende Lokalisationslehre der Grosshirnrinde in ihren Prinzipien dargestellt auf Grund des Zellenbaues.* Leipzig: Barth; 1909.

Caine, E. D., and Joynt, R. J. Neuropsychiatry . . . again. *Arch Neurol* 1986; 43: 325–327.

Churchland, P. S. *Neurophilosophy: Toward a Unified Science of the Mind-Brain.* Cambridge: MIT Press; 1986.

Corsi, P., ed. *The Enchanted Loom: Chapters in the History of Neuroscience.* New York: Oxford University Press; 1991.

Crevier, D. *AI: The Tumultuous History of the Search for Artificial Intelligence.* New York: Basic Books; 1993.

Cummings, J. L. Frontal-subcortical circuits and human behavior. *Arch Neurol* 1993; 50: 873–880.

Cummings, J. L., and Hegarty, A. Neurology, psychiatry, and neuropsychiatry. *Neurology* 1994; 44: 209–213.

Cummings, J. L., and Mega, M. S. *Neuropsychiatry and Behavioral Neuroscience.* New York: Oxford University Press; 2003.

Damasio, A. R. Behavioral neurology: theory and practice. *Semin Neurol* 1984; 4: 117–119.

Damasio, H., and Damasio, A. R. *Lesion Analysis in Neuropsychology.* New York: Oxford University Press; 1989.

Dennett, D. C. *Consciousness Explained.* Boston: Little, Brown and Company; 1991.

Descartes, R. *Discourse on Method.* Paris: n.p.; 1637. Trans. Sutcliffe: F. E. London, Penguin; 1968.

D'Esposito, M. Functional neuroimaging of cognition. *Semin Neurol* 2000; 20: 487–498.

Duffy, C. J. The legacy of association cortex. *Neurology* 1984; 34: 192–197.

Filley, C. M. *The Behavioral Neurology of White Matter.* New York: Oxford University Press; 2001.

Friehs, G. M, Zerris, V. A., Ojakangas, C. L., et al. Brain-machine and brain-computer interfaces. *Stroke* 2004; 35 (11 Suppl 1): 2702–2705.

Gall, F. J., and Spurzheim, J. K. *Anatomie et physiologie de système nerveux en général et du cerveau en particular*. Paris: Schoell; 1810–1818.

Gardner, H. *The Mind's New Science: A History of the Cognitive Revolution*. New York: Basic Books; 1987.

Geschwind, N. The borderland of neurology and psychiatry: some common misconceptions. In: Benson, D. F., and Blumer, D., eds. *Psychiatric Aspects of Neurologic Disease*, vol. 1. New York: Grune and Stratton; 1975: 1–8.

―――. Brain disease and the mechanisms of mind. In: Coen, C. W., ed. *Functions of the Brain*. Oxford: Clarendon Press; 1985: 160–180.

―――. Disconnexion syndromes in animals and man. *Brain* 1965; 88: 237–294, 585–644.

Goldstein, K. *Language and Language Disturbances*. New York: Grune and Stratton; 1948.

Goodkin, D. E., Rudick, R. A., and Ross, S. J. The use of brain magnetic resonance imaging in multiple sclerosis. *Arch Neurol* 1994; 51: 505–516.

Gorman, D. G., and Unützer, J. Brodmann's "missing" numbers. *Neurology* 1993; 43: 226–227.

Horgan, J. Can science explain consciousness? *Sci Am* 1994; 271: 88–94.

Hubel, D. H., and Wiesel, T. N. Functional architecture of macaque monkey visual cortex. *Proc R Soc Lond B* 1977; 198: 1–59.

James, W. *The Principles of Psychology*. New York: Henry Holt and Company; 1890.

Kandel, E. R. Genes, nerve cells, and the remembrance of things past. *J Neuropsychiatry Clin Neurosci* 1989; 1: 103–125.

―――. Psychotherapy and the single synapse. *N Engl J Med* 1979; 301: 1028–1037.

Kandel, E. R., Schwartz, J. H., and Jessell, T. M., eds. *Principles of Neural Science*. 4th ed. New York: McGraw-Hill; 2000.

Lashley, K. S. Studies of cerebral function in learning VII: the relation between cerebral mass, learning and retention. *J Comp Neurol* 1926; 41: 1–58.

Lezak, M. D., Howieson, D. B., Loring, D. W., et al. *Neuropsychological Assessment*. 4th ed. New York: Oxford University Press; 2004.

Lishman, W. A. What is neuropsychiatry? *J Neurol Neurosurg Psychiatry* 1992; 55: 983–985.

Luria, A. R. *Higher Cortical Functions in Man*. New York: Consultants Bureau; 1980.

Mesulam M.-M. Behavioral neuroanatomy: large-scale networks, association cortex, frontal systems, the limbic system, and hemispheric specializations. In: Mesulam M.-M. *Principles of Behavioral and Cognitive Neurology*. 2nd ed. New York: Oxford University Press; 2000: 1–120.

―――. Large-scale neurocognitive networks and distributed processing for attention, language, and memory. *Ann Neurol* 1990; 28: 597–613.

Mountcastle, V. B. An organizing principle for cerebral function: the unit module and the distributed system. In: Edelman, G. M., and Mountcastle, V. B., eds. *The Mindful Brain*. Cambridge: MIT Press; 1978: 7–50.

Naeser, M. A., and Hayward, R. W. Lesion localization in aphasia with cranial computed tomography and the Boston Diagnostic Aphasia Exam. *Neurology* 1978; 28: 545–551.

Nauta, W.J.H., and Fiertag, M. *Fundamental Neuroanatomy*. New York: W. H. Freeman; 1986.

Nolte, J. *The Human Brain: An Introduction to Its Functional Anatomy*. 5th ed. St. Louis: Mosby; 2002.

Parent, A. *Carpenter's Human Neuroanatomy*. 9th ed. Baltimore: Williams and Wilkins; 1996.

Penfield, W. *The Mystery of the Mind*. Princeton, NJ: Princeton University Press; 1975.

Popper, K. F., and Eccles, J. C. *The Self and Its Brain*. Berlin: Springer International; 1977.

Posner, M. I. Seeing the mind. *Science* 1993; 262: 673–674.

Price B. H., Adams, R. D., and Coyle, J. T. Neurology and psychiatry: closing the great divide. *Neurology* 2000; 54: 8–14.

Prichard, J. W., and Cummings, J. L. The insistent call from functional MRI. *Neurology* 1997; 48: 797–800.

Raichle, M. E. Visualizing the mind. *Sci Am* 1994; 270: 58–64.

Roland, P. E. *Brain Activation*. New York: Wiley-Liss; 1993.

Rudkin, T. M., and Arnold, D. L. Proton magnetic resonance spectroscopy for the diagnosis and management of cerebral disorders. *Arch Neurol* 1999; 56: 919–926.

Ryle, G. *The Concept of Mind*. Chicago: University of Chicago Press; 1949.

Saver, J. L., and Rabin, J. The neural substrates of religious experience. *J Neuropsychiatry Clin Neurosci* 1997; 9: 498–510.

Schoenemann P. T., Sheehan, M. J., and Glotzer, L. D. Prefrontal white matter volume is disproportionately larger in humans than in other primates. *Nat Neurosci* 2005; 8: 242–252.

Searle, J. R. *Mind: A Brief Introduction*. New York: Oxford University Press; 2004.

———. *Minds, Brains and Science*. Cambridge, MA: Harvard University Press; 1984.

Skinner, B. F. *Beyond Freedom and Dignity*. New York: Alfred A. Knopf; 1971.

Trimble, M. R. Neuropsychiatry or behavioural neurology. *Neuropsychiatry Neuropsychol Behav Neurol* 1993; 6: 60–69.

Turner, R. Magnetic resonance imaging of brain function. *Ann Neurol* 1994; 35: 637–638.

Wright, I. C., McGuire, P. K., Poline, J. B., et al. A voxel-based method for the statistical analysis of gray and white matter density applied to schizophrenia. *Neuroimage* 1995; 2: 244–252.

Young, J. Z. *Philosophy and the Brain*. New York: Oxford University Press; 1987.

Yudofsky, S. C., and Hales, R. E. The reemergence of neuropsychiatry: definition and direction. *J Neuropsychiatry Clin Neurosci* 1989; 1: 1–6.

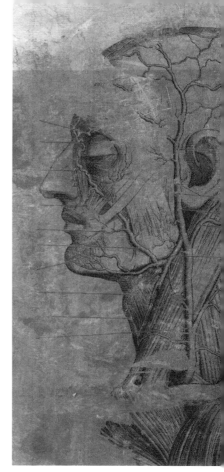

CHAPTER TWO

MENTAL STATUS
EVALUATION

A central goal of the clinical interaction between physician and patient is the gathering of information that will contribute to the patient's proper evaluation and care. In the case of neurobehavioral disorders, deficits due to brain dysfunction can be recognized only after a thorough assessment of behavior has been completed. Many of the clinical data, of course, are elicited with a medical history, physical examination, elemental neurological examination, and appropriate laboratory, neuroimaging, and neuropsychological testing, but the assessment of mental status offers a particularly revealing, if challenging, portion of the clinical encounter. The mental status evaluation is in fact a foundation of behavioral neurology (Strub and Black 1993; Cummings and Mega 2003).

HISTORY AND INTERVIEW

Experience in clinical medicine teaches that every disorder has a history. That is, the symptoms endorsed by a patient—or, as is true of many neurobehavioral disorders, described by informed observers—are the key phenomena of the clinical problem and highlight its beginning and evolution until the clinical encounter takes place. History-taking is a crucial component of clinical care, and seasoned clinicians can often be confident of a diagnosis solely on the basis of focused and informed elicitation of the history. The mental status examination is of course necessary and often discloses much additional useful detail, but in many cases this examination is confirmatory of what the clinician is already suspecting from the history. This principle holds as true for neurobehavioral disorders as for any others in medicine.

As a first step, much can be learned by simple observation of the patient during the initial greetings and history-taking. In some cases, cognitive dysfunction is immediately evident by the fact that the individual must be accompanied by a spouse or caregiver because the capacity to arrange for and arrive at a clinic visit is already compromised. The general appearance of the patient gives information about the degree of self-care and the quality of comportment, as well as evidence of cranial trauma or surgery, lateralized or focal weakness, movement disorders, gait disturbance, distractibility, agitation, apathy, catatonia, compulsive acts, anxiety, and hyperactivity. Abnormalities such as these may be obvious or subtle, and a wide range of physical signs can reflect neurobehavioral dysfunction. During the interview, a change in the patient's personality may become apparent as the details of the illness unfold. Anomalous thought characteristics such as loosened associations, tangentiality, flight of ideas, and grandiosity often become apparent in conversation. Delusions, the most notable disorders of thought content, may emerge. Abnormal perceptual phenomena, including hallucinations and illusions, may be obvious or easily elicited by appropriate questioning. Detection of disturbances such as these may suggest a behavioral disorder not clearly related to structural brain disease (a "psychiatric" disorder), and the remainder of the evaluation can be adjusted accordingly. Conversion, somatization, and other somatoform disorders are suggested by the absence of structural or metabolic disease in spite of worrisome neurologic symptoms and extensive evaluation, and in rare cases, malingering can be suspected if evidence of attempted deception is uncovered (American Psychiatric Association 1994).

It is good to remember, however, that observations suggesting psychiatric disease may also reflect a neurologic illness (Cummings and Mega 2003). *Delusions*, or fixed false beliefs, have been reported in a great variety of neurologic conditions (Cummings 1985) and appear to be more frequent when the pathologic process affects the limbic system (Cummings 1992). Paranoid delusions are common in Alzheimer's disease (AD), for example, and include the *Othello syndrome*, or delusional jealousy, in which a person harbors the unjustified conviction of a spouse's infidelity (Shepard 1961). *Erotomania*, or de Clérambault's syndrome, is a delusion seen mostly in women who believe that an older, influential man is in love with them despite all evidence to the contrary (Anderson, Camp, and Filley 1998). Three related delusional disorders are the *Capgras syndrome* (Alexander, Stuss, and Benson 1979), the *Fregoli syndrome* (Feinberg et al. 1999), and *reduplicative paramnesia* (Benson, Gardner, and Meadows 1976; Filley and Jarvis 1987). All are misidentification phenomena associated with a combination of bifrontal and right hemisphere pathology. In the Capgras syndrome, the patient believes that a family member or friend has been replaced by an impostor; the Fregoli syndrome, in contrast, features the belief that a persecutor is taking on the appearance of a stranger; and in reduplicative paramnesia, the delusion exists that a familiar place has been relocated to a different site.

Hallucinations, sensory experiences without external stimulation, may occur in visual, auditory, tactile, olfactory, or gustatory modalities (Siegel 1977). Some general rules guide the history-taking. Visual hallucinations generally suggest neurologic involvement, whereas auditory hallucinations imply psychiatric disease. Three settings predispose to visual hallucinations: perceptual release due to visual system disease (Lepore 1990), ictal discharge (King and Marsan 1977), and toxic-metabolic disorders (Lipowski 1980). Two types of release visual hallucination are the *Charles Bonnet syndrome*, visual hallucinations due to encroaching blindness from ocular pathology (Damas-Mora, Skelton-Robinson, and Jenner 1982), and *palinopsia*, recurrence of a visual image after removal of the stimulus, associated with right parietooccipital lesions (Stagno and Gates 1991). Visual hallucinations are more common with right than left hemisphere lesions (Lessell 1975; Hecaen and Albert 1978), and, in general, unformed visual hallucinations suggest occipital lobe lesions whereas formed images imply temporal lobe pathology (Cummings and Mega 2003). A peculiar type of visual hallucination is *peduncular hallucinosis*, which refers to the presence of formed hallucinations, often of people or animals, that are interpreted as benign or even entertaining; midbrain lesions in the vicinity of the cerebral peduncle are thought to be responsible (Dunn, Weisberg, and Nadell 1983). Auditory hallucinations are very common in schizophrenia but may also occur in *schizophreniform psychosis* (Chapter 9) and *alcoholic hallucinosis* (Victor and Hope 1958). *Palinacousis* is analogous to palinopsia, and the recurrent auditory experiences usually relate to a temporal lobe lesion (Jacobs et al. 1973). Tactile hallucinations occur mainly in psychiatric disorders but can also appear in toxic-metabolic disorders (Berrios 1982); *formications*, the feeling of insects crawling in the skin, are common in drug withdrawal states. Olfactory, and rarely gustatory, hallucinations can be experienced with temporal lobe seizures (Daly 1975).

Illusions are misperceptions of external stimuli. They occur usually in the visual realm and differ from hallucinations in that they do not arise spontaneously but are provoked by visual input. Common illusions include *micropsia* (the appearance of decreased size), *macropsia* (the appearance of increased size), and *metamorphopsia* (a change in shape or form). The most common settings favoring the occurrence of illusions are temporal lobe seizures and classic migraine; in both, an electrical discharge of the relevant cerebral cortical region is thought to be responsible (Hecaen and Albert 1978).

MENTAL STATUS EXAMINATION

After the history-taking, the mental status examination is then carried out in a manner analogous to the remainder of the neurological examination. The objective is to systematically assess specific functions so that a syndrome diagnosis can be made; then, alterations in underlying anatomy and physiology can be deduced and treatment planned (Strub and Black 1993; Cummings and Mega 2003). It is therefore obvious that some appreciation of the neurobiological foundations of these functions is essential to the examiner. What follows is a comprehensive review of the mental status examination, which provides the organizational framework for considering the syndromes described in this book. This examination can be conveniently divided into six broad categories: arousal, attention, and motivation; memory; language; visuospatial function; executive function; and emotion and personality (Table 2.1). The domains listed in Table 2.1 organize the diagnostic approach to major neurobehavioral syndromes and should all be tested in more or less detail during neurobehavioral evaluation; additional comments on the evaluation of specific syndromes can be found in the appropriate chapters.

Certain principles should be remembered in conducting these examinations. First, clinical tests seldom assess a single neurobehavioral domain in isolation, although they can be fairly specific; thus, the data collected must be analyzed as a whole to detect the salient deficits. Related to this is that findings of the elemental neurological examination—meaning the testing of cranial nerves, motor system, coordination, gait, reflexes, and sensation—can often provide essential localizing information to support a neurobehavioral diagnosis; for example, some degree of right hemiparesis is typically expected in Broca's aphasia and helps confirm this diagnosis. Another point is that many aspects of a patient's performance are significantly influenced by age, education, cultural background, and life experiences, necessitating considerable sensitivity and flexibility on the part of the examiner. Finally, it should be stressed that keen observation is of paramount importance; neurology, particularly behavioral neurology, remains heavily dependent on the clinician's skill in detecting the subtle, often nonverbal, manifestations of central nervous system function and dysfunction. Even powerful new neuroimaging techniques cannot replace the careful and sensitive evaluation of an experienced clinician.

TABLE 2.1. Mental status examination

I. Arousal, Attention, and Motivation	**IV. Visuospatial Ability**
A. Arousal	A. Drawing
1. Hyperarousal	1. Two- and three-dimensional objects
2. Hypoarousal	2. Clock with hands at 11:10
B. Attention	B. Directed attention
1. Digit span	1. Double simultaneous stimulation
2. Serial sevens	2. Line cancellation
3. Random letter test	3. Line bisection
C. Motivation	C. Spatial orientation
1. Cooperation	D. Dressing
2. Perseverance	E. Gnosis
II. Memory	**V. Executive Function**
A. Immediate	A. Abstraction
B. Recent	B. Word-list generation
1. Verbal	C. Motor programming
2. Visual	D. Alternating sequences
C. Remote	E. Insight
	F. Judgment
III. Language	**VI. Emotion and Personality**
A. Propositional language	A. Mood
1. Spontaneous speech	1. Inquiry to patient
2. Auditory comprehension	2. Inquiries to family, caregivers
3. Repetition	B. Affect
4. Naming	1. Facial expression
5. Reading	2. Tone of voice
6. Writing	3. Latency of response
B. Prosody	4. Gesture and body posture
C. Praxis	C. Comportment
D. Calculations	

Arousal, Attention, and Motivation

The formal mental status examination logically begins with a consideration of the fundamental functions of arousal, attention, and motivation. Assessment of these phenomena is a necessary first step in judging mentation because if any one is sufficiently compromised, the remainder of the examination is uninformative. Individuals who are not fully awake, who do not adequately pay attention, or who fail to put forth satisfactory effort cannot be formally assessed in neurobehavioral terms, although of course much can still be learned in other ways about their mental state.

Arousal is a physiologic concept describing the wakefulness of the individual. It is useful to regard arousal in terms of the level of consciousness, which is best viewed as dependent on the integrity of the ascending reticular activating system (ARAS) in the brainstem and diencephalon (Posner et al. 2007). In contrast, the *content of consciousness*—the sum of all the operations of cognition and emotion—is mediated primarily by the cerebral hemispheres (Posner et al. 2007). It is evident that an appropriate degree of arousal is required for the engagement of cognitive and emotional processes; in neuroanatomic terms, the ARAS must be functional for the hemispheres to operate effectively.

Disorders of arousal are identified by simple recognition of abnormal states, be they those of hyperarousal (restlessness, agitation, delirium) or hypoarousal (drowsiness, lethargy, stupor, coma). The use of many of these terms is not standardized, and if doubt or the possibility of poor communication exists, a wise course is simply to describe a patient's level of consciousness with specific behavioral observations rather than attach to it an ambiguous label.

Attention is a more complicated phenomenon, and its many different definitions by various authorities have led to considerable confusion. Nevertheless, inattention in a neurological sense has a variety of important diagnostic and therapeutic implications. For the purposes of this book, attention will be considered in light of the many disorders that have been noted to lead to some aspect of attentional dysfunction (Posner et al. 2007). This approach will then permit a practical, albeit preliminary, formulation of how attentional networks in the brain are organized.

The term *alertness* is frequently used in clinical practice to describe attention, but more specifically, three aspects of attention can be distinguished and shown to have neurobehavioral implications (Filley 2002). First is the category of *selective attention*, a capacity that permits the focusing of awareness on biologically relevant stimuli in the environment. This crucial ability is reliably judged by the digit span test, in which the normal performance is repetition of 7 ± 2 numbers forward and 5 ± 1 numbers backward (Cummings and Mega 2003). Second on this list is *sustained attention*, also called vigilance or concentration, which involves selective attention to a task longer than the few seconds necessary for the digit span. Sustained attention can be evaluated by the classic serial sevens test, counting backward from 100 by 7s, or by the random letter test, in which the patient indicates each time the examiner says the letter *A* in a long random list of letters (Strub and Black 1993). Finally, the variant known as *directed attention* is a lateralized function that is organized in such a fashion that the right hemisphere attends to the left hemispace and vice versa. Dysfunction of this system—resulting in the phenomenon of hemineglect or simply neglect—will be considered below.

Motivation refers to the level of effort a patient puts forth during the examination and is subjectively but adequately assessed by noting the degree of cooperation and perseverance. The patient's underlying personality clearly plays a major role in motivation, but this aspect of behavior is also dependent on the integrity of frontal, subcortical, and limbic structures (Mesulam 2000). Amotivational states often imply bilateral frontal lobe dysfunction, and depression is another potential cause. Malingering with symptom magnification and other psychiatric conditions need also to be considered in this context (American Psychiatric Association 1994).

Memory

The word *memory* has been employed to refer to a group of related psychological functions, and the study of memory has taken many different approaches depending on the discipline undertaking its investigation. One source of continuing confusion is the distinction between short-term and long-term memory, a dichotomy used in the psychological literature to distinguish between temporary and permanent storage of memory. This difference is indeed meaningful, but does long-term memory refer to new learning within minutes or to stable memories over months or years? Clinicians have tended to divide long-term memory into the categories of recent and remote, while retaining a form of short-term memory in the category designated as immediate (Table 2.1). For clinical purposes, therefore, it is most useful to adhere to a tradition in neurology that considers memory under the headings of immediate, recent, and remote (Strub and Black 1993; Cummings and Mega 2003).

Immediate memory is intimately related to selective attention, discussed above, which enables the performance of tasks such as recalling a telephone number long enough to use it correctly without the need to write it down. An older formulation of this general phenomenon is *primary memory* (James 1890). More recently, another roughly equivalent term for this capacity is *working memory* (Baddeley 1992). Working memory is generally considered the capacity to hold information in awareness long enough to perform a cognitive operation before the information is discarded. Whatever terminology is used, this kind of memory is adequately tested with the digit span test. Alternatively, counting backward from 10, or reciting the months of the year or the days of the week in reverse, is a satisfactory procedure.

Recent memory refers to the ability to accomplish new learning. Information retained for a period of minutes to hours is regarded as being held in recent memory. Testing of recent memory involves asking the patient to remember three or four unrelated words for 5 to 10 minutes, during which time other components of the evaluation are carried out to prevent mental rehearsal (Strub and Black 1993). If this task is failed, provision of a semantic cue (such as the category for the forgotten item) or a phonemic cue (the first phoneme of the word) may help certain patients, and offering a list of words from which the correct one can be recognized can assist others (Cummings and Mega 2003). Individuals whose performance improves with these procedures generally have more difficulty with retrieval than with encoding and may have primarily frontal-subcortical neuropathology or a psychiatric disorder. As it is known that left- and right-sided lesions may cause selective verbal and visual memory deficits, respectively, it is often worthwhile to use the same format with pictures, designs, or objects in the room. The Three Words–Three Shapes test, for example, allows simultaneous evaluation of verbal and nonverbal recent memory (Weintraub 2000). Other tests of recent memory involve orientation to locale and date, as well as details of the present illness or current political or sports events. Procedural memory, a term referring to skill learning, is not routinely tested in the office or clinic but can be studied in the cognitive neuropsychology laboratory (Weintraub 2000) and has important implications that will be discussed in Chapter 12.

Remote memory, commonly known as knowledge, involves the storage of material learned days, months, or years past and is tested by asking about significant occasions in the patient's life or important historical events. This component of memory is assumed to depend on diffuse, probably redundant representation of information in the neocortex. As such, remote memory is typically better preserved than recent memory in neurologic disease, a feature strikingly apparent in many cases of early dementia. Once a memory is lodged in the neocortex, it is relatively resistant to neuropathology, but this feature is of little help to the many patients who present with memory disorders that feature problems with the initial deposition of information.

Language

Language is the verbal or written representation of thought that permits symbolic communication with other individuals. It is to be distinguished from *speech*, the motor capacity for the articulation used in the production of oral language, and *voice*, the laryngeal function of coordinating phonation and respiration. Disorders of speech (dysarthrias) and voice (dysphonias) may or may not be associated with disorders of language (aphasias, or, as some authorities prefer, dysphasias) and require separate consideration.

Propositional language refers to the word choice and word order of language and differs from the paralinguistic aspect of language known as prosody, which imparts affect and emotion to language by such qualities as pitch, volume, intonation, melody, tempo, stress, and timing (Ross 2000). Propositional language deficits are those that lead to aphasia, the testing for which is a crucial task for clinicians. In the process of this testing, six aspects of language should always be sampled if the patient is capable of performing the requested tasks (Goodglass and Kaplan 1983).

Spontaneous speech is assessed during conversation in the history-taking, and the examiner should mainly assess the patient's fluency or nonfluency. The latter feature is characterized by reduced phrase length (a maximum of five words or less between pauses), agrammatism (incorrect grammatical structure), impaired linguistic prosody (abnormal rhythm and stress), and articulatory struggle (dysarthria). Skill is often needed for making the distinction between fluency and nonfluency, and the two may sometimes be more appropriately seen as lying on a continuum rather than as a dichotomy (Hillis 2007). *Auditory comprehension* is probed by the presentation of increasingly difficult directions: pointing to one's body parts such as knuckles or earlobes; pointing to objects in the room; pointing to two, three, or four objects in sequence; pointing to objects functionally described (e.g., "Show me the device I have on to tell time"); answering sentence-length yes-no questions (e.g., "Am I wearing a hat?", "Is a fork good for eating soup?"); and answering questions that employ more difficult grammar and syntax (e.g., "A lion was killed by a tiger; which animal died?"). *Repetition* is tested by having the patient repeat single content words (e.g., "house," "baseball"), sentences with content words (e.g., "The train entered the station"), and sentences with many small functor words (e.g., "He is the one who did it"). *Naming* is done by the method of confrontation, so that the patient is asked to name visually presented objects or body parts. Failure to name high-frequency (common) items suggests a more severe naming deficit than a problem naming low-frequency (uncommon) items alone. *Reading* is tested by having the patient read aloud and then silently for comprehension; both are performed using letters, words, sentences, and paragraphs. *Writing* involves the patient producing a signature (a highly resistant, overlearned skill), writing to dictation (words, short sentences), and writing a narrative sentence about some familiar topic, such as the current weather conditions. Reading and writing, of course, are only testable if the patient is literate in the language used for the examination.

In addition to this basic core of tests, the presence of *paraphasias* should be noted. These are errors made in speech characterized by letter or word substitutions and are classified as phonemic or literal (e.g., "bree" for "tree"), semantic or verbal (e.g., "house" for "tree"), or neologistic (a neologism is a meaningless word such as "stribenlug"). Paraphasic speech is characteristic of all aphasias, but it is easier to detect in fluent aphasias in which there is more abundant speech production. The term *jargon aphasia* refers to severe fluent aphasia with heavily paraphasic speech that is difficult or impossible to follow. In selected patients, other procedures may be useful. Evaluation of singing can demonstrate musical ability that may be advantageous in the rehabilitation of nonfluent aphasics (Chapter 8). The finding that serial speech (days of the week, the pledge of allegiance) is preserved assists in the diagnosis of the transcortical aphasias. Spelling errors can accompany the abnormal written output of aphasia (agraphia) or be associated with developmental reading syndromes.

Prosody, as mentioned above, is a paralinguistic aspect of language referring to the inflection of propositional language (Ross 2000). Without prosody, language can express routine declarative and interrogative utterances but not the richness of communication conferred by affect and emotion—including such aspects as delight, sadness, humor, surprise, innuendo, and the like. The right hemisphere is thought responsible for prosody, in keeping with its general affiliation with many dimensions of emotion (Ross 2000). Aprosody is often obvious in conversation when an affected patient speaks in an uninflected, monotonous tone, and more subtle forms can be disclosed by formal testing, as will be discussed in Chapter 8.

It is often appropriate after the assessment of language to examine for *apraxia*, a disorder of the ability to carry out learned motor activity. Apraxia often accompanies aphasia and may exacerbate linguistic deficits from which aphasic patients already suffer. Furthermore, apraxia provides localizing information in addition to that revealed by language disturbances. The fundamental procedure involves asking patients to carry out skilled movements in the limbs, buccofacial region, and axial musculature. Apraxia is discussed as a disorder of higher motor function in Chapter 6.

Finally, *calculations* can be logically tested at this point in the examination. The capacity to perform arithmetic operations is closely related to language, and many patients with aphasia also have some degree of acalculia. Other related functions can be assessed as well, and the testing for *Gerstmann's syndrome* (acalculia, agraphia, right-left disorientation, and finger agnosia) fits conveniently here into the examination (Strub and Black 1993). After a writing sample is obtained, as described above, calculating ability is then probed using simple and more challenging arithmetic problems. Next, the patient is asked to identify his or her right and left hands and those of the examiner. Finally, recognition of fingers (thumb, index finger, middle finger, ring finger, and little finger) on the patient and the examiner is tested. This peculiar combination of deficits often localizes a lesion to the left angular gyrus. Finger agnosia is one setting in which the word agnosia finds use, but the classic agnosias refer more properly to modality-specific recognition deficits. Agnosia as a disorder of higher sensory function can be diagnosed when higher-order deficits in vision, audition, and tactile sensation are detected; the agnosias will be considered in Chapter 7.

Visuospatial Ability

This broad category includes the ability to attend to visually presented material, analyze it, remember it, and represent it accurately by means of integrated motor output. Traditionally, these abilities are assigned to the right hemisphere, and a large body of data supports the central role of the right hemisphere in visuospatial processing (Mesulam 2000; Cummings and Mega 2003). Tests probing various aspects of visuospatial ability offer the most convenient means of surveying the neurobehavioral integrity of the right hemisphere. The right parietal lobe, however, is considered the most important single region for the performance of visuospatial tasks.

The most familiar and well-established test of visuospatial function is *drawing*, which evaluates the patient's constructional ability (Strub and Black 1993). First, drawing to the examiner's direction is conducted: simple geometric shapes (e.g., square, triangle, cross), three-dimensional shapes (e.g., cube, house), and flowers, such as a daisy, or a clock face. Then it is often helpful to ask the patient to copy the examiner's drawings: objects, if any, that were previously failed, and any new ones that may be instructive. Regardless of which drawing task is employed, abnormalities in patient performance suggest neurologic disease since constructional abilities are usually spared in idiopathic psychiatric disorders (Cummings and Mega 2003). In general, right frontal lesions disrupt drawing more than copying, whereas right parietal lesions exert a greater effect on copying than drawing (Cummings and Mega 2003). The clock face with the hands set at 11:10 is particularly helpful because it may disclose hemineglect or stimulus-bound behavior in addition to visuospatial disturbance (Cummings and Mega 2003). Errors typically made by patients with right hemisphere lesions include left hemineglect (see below), loss of perspective, and a tendency to work from right to left; lesions of the left hemisphere typically have less impact on constructional ability, but affected patients may display simplification, omissions, and loss of internal detail (Cummings and Mega 2003). Recent evidence has found that visuospatial errors on clock drawing implicate right parietal lesions whereas time-setting errors suggest left frontoparietal damage (Tranel et al. 2008).

Equally instructive tests of visuospatial function assess the domain of *directed attention*. This is a lateralized function organized in such a fashion that the right hemisphere attends to the left hemispace and vice versa. Hemiattentional disorders, most importantly *hemineglect*—often referred to simply as *neglect*—are sought by testing for extinction to double simultaneous stimulation in the tactile, visual, and auditory modalities. Affected patients will fail to detect stimuli that are simultaneously applied to both sides of the body but have no difficulty with unilateral stimulation. Neglect is much more common on the left side of the body than the right, and the demonstration of left neglect should prompt a particularly assiduous search for other signs of right hemisphere dysfunction. Other useful ways to search for neglect are tests of line cancellation (asking the patient to cross out all the randomly placed lines on a sheet of paper) and line bisection (asking for a mark to be made exactly in the middle of each of several horizontal lines of differing length).

Spatial orientation requires a set of skills closely related to those supporting constructional ability (Strub and Black 1993) and refers to the ability to perform such tasks as recognizing one's physical location or reading a map correctly.

Often deficits in this domain will be apparent from the history, as in a person who can no longer drive confidently home from a familiar site, or from drawing errors made earlier in the examination. Further testing can be accomplished by having the patient place major cities or other sites on a sketch of a well-known country. The patient may then be asked how one would travel to some familiar place from the hospital or clinic, or how to get to the nurses' station or the receptionist's desk.

Dressing is an important activity of daily living that can be significantly disrupted by right hemisphere lesions. The capacity to successfully don an item of clothing requires an understanding of the dimensions and structure of one's body as well as the garment itself, and a disorder of the body schema is common with damage to the right parietal lobe (Cummings and Mega 2003). Dressing can be judged by simply requesting the patient to put on a jacket, shirt, or shoe, and the task can be made more challenging by, for example, pulling out one arm of the jacket or shirt, or presenting the right shoe to the left foot. If so-called dressing apraxia is found, other information pertaining to right parietal dysfunction can be sought.

Finally, the domain of *gnosis* can be conveniently assessed at this point in the examination. At the most basic level, gnosis means the recognition of sensory stimuli and can be meaningfully considered in the visual, auditory, and tactile modalities. Disorders of gnosis are complex and relatively rare, since deficits in recognition are often commingled with deficits in primary sensation or other cognitive realms, but careful testing can disclose these syndromes when they occur. A key point is that agnosia is modality-specific, so that a deficit must be confined to a given sensory channel. *Visual agnosia*, the most prominent member of this group, can be suspected by observing that a patient cannot recognize an object presented visually but can do so when the object is presented in another modality; hence, for example, a set of keys is not recognized when seen by the patient but is easily identified when jingled by the examiner or placed in the patient's hand. *Auditory agnosia* is closely related to aphasia, and *tactile agnosia* is a sign of parietal lobe dysfunction. In Chapter 7 all of the agnosias will be covered in detail.

Executive Function

Recent years have witnessed a growing consensus that *executive function* deserves special attention as a central domain of human behavior (Stuss and Benson 1986; Miller and Cummings 1999; Cummings and Mega 2003; Filley 2009). Distinct from attention, memory, language, praxis, gnosis, and visuospatial ability, the abilities subsumed under the rubric of executive function are among the brain's "highest" functions and in many respects exert a supervisory role in human consciousness. Executive function, however, is perhaps more diffi-cult to evaluate in neurobehavioral terms since it involves complicated cognitive operations that often defy quantitation. Nevertheless, the prominence of frontal systems in executive function is well established, and identification of executive dysfunction has important localizing value (Stuss and Benson 1986; Miller and Cummings 1999; Cummings and Mega 2003; Filley 2009). The examination of executive function attempts to assess the patient's capacity to plan, sustain, and monitor cognitive activities while averting distractions and adapting to unex-pected situations. These capacities clearly implicate the frontal lobes and their connections to subcortical and more posterior regions.

Tests of executive function frequently given include tasks involving *abstrac-tion*. The interpretation of idioms (e.g., "a loud tie," "a heavy heart") and prov-erbs (e.g., "Don't cry over spilled milk"; "You can't tell a book by its cover"), which tap these kinds of reasoning skills, and recognition of similarities (e.g., table and chair, coal and paper) assess metaphorical capacity in a rudimentary fashion. *Word-list generation* is a verbal fluency task requiring the patient to list aloud in 60 seconds as many words as possible within a given category, such as animals or fruits and vegetables (normal is 18 or more per category), or as many words as possible that begin with a specific letter such as *F*, *A*, or *S* (normal is 12 or more per letter) (Strub and Black 1993). *Motor programming* tasks were first developed by the Russian neuropsychologist A. R. Luria and require the patient to execute a series of three hand movements—fist, edge, and palm—in sequence (Cummings and Mega 2003). *Alternating sequence* tasks are similar, and the patient is asked to copy a continuous series of alternating *m*'s and *n*'s in cursive, or alternating squares and triangles; after a short sequence to copy, the patient continues to the end of the page, and perseveration may be demonstrated if the alternating sequence is not maintained (Cummings and Mega 2003). Much can be learned about emotional adjustment by inquiring into the patient's *insight* into the illness—how it will affect him or her and the family or other caregivers. In addition, simple questions regarding *judgment* (e.g., "What would you do in a crowded theater if you smelled smoke?") may reveal evidence of disturbed social and ethical behavior.

Finally, the functions that are truly the highest in the human behavioral repertoire—creativity, altruism, love, and the like—prove difficult to assess sub-jectively, much less quantify, and clinical disturbances in these attributes may be harder to characterize. In a remarkable development stimulated largely by emerging neuroimaging technologies, neuroscientists are beginning to explore the neurobiological foundations of these highly subjective domains in ways that were inconceivable only a few decades ago; there are now reasonable data bear-ing on the cerebral organization of creative behavior (Austin 2003; Heilman 2005), social cooperation (Rilling et al. 2002), and even love itself (Zeki 2007).

In the clinic, however, one can realistically expect to uncover only a general sense of alteration in these capacities after some time is spent with a patient. Sometimes changes in work performance or social competence, as recounted by a colleague, supervisor, or spouse, may be the sole area of compromise reflecting a disturbance in highly evolved behaviors. Information from family or friends may be most helpful in this regard, as it is often the case that the erosion of a previous virtue may not be apparent to an affected person precisely because the lesion causing this deficit also produces loss of insight. These considerations lead directly into the final category of the mental status examination, to be considered next.

Emotion and Personality

The concepts of emotion and personality are familiar to most people but defy precise definition. Much can be written about these singular human attributes, but for the purposes of this chapter, *emotion* will be considered the feeling state of an individual, and *personality* the characteristic repertoire of traits that determine everyday interactions with others. Our main purpose in this section will be to distinguish these aspects of behavior from those that have been discussed above and are typically considered "cognitive." Derived from the Latin verb *cognoscere*, meaning "to know," cognition is used in this book to refer broadly to the everyday notion of "thinking" and includes aspects of behavior associated with attention, memory, language, praxis, visuospatial skills, gnosis, and executive function. The localization of cognitive functions is relatively well understood and is the main topic of this book. The neuroanatomical organization of emotion, on the other hand, is not well defined. The limbic system clearly plays a pivotal role, mediating the drives or instincts that assist in the attainment of fundamental biological needs. Limbic structures, especially the amygdala and cingulate gyrus, participate in the experience of emotion, and the hypothalamus is responsible for directing the autonomic and endocrine effectors of emotion. The activity of these systems is then modified by the influence of neocortical areas, especially of the frontal lobes, which act, as reviewed above, to modulate basic drives into acceptable social patterns. Also important in emotional behavior is the right hemisphere, which acts in little-understood ways to confer elements such as interpersonal competence, prosodic and musical skills, and humor upon the behavioral profile. Emotional disorders may thus result from damage to a variety of areas.

It can be argued that all higher functions—including the emotions—are in fact fundamentally cognitive, and recently a body of literature has appeared dealing with the notion of *social cognition* (Frith 2008; Adolphs 2009). This term reflects a growing appreciation of the brain's mediation of social aspects of human behavior such as empathy, compassion, and morality. These traits, which logically appear to be mediated by frontal and temporal lobe structures, merit inclusion as neurobiological processes as much as cognitive functions such as memory and language. For practical purposes, however, a general distinction between cognition and emotion still proves convenient. In clinical practice, despite the gradual rapprochement between neurology and psychiatry now under way, the care of people with "cognitive" dysfunction is still assigned to the former specialty while the care of those with "emotional" disorders is referred to the latter. We will therefore proceed with this dichotomy as an organizing principle, recognizing that all domains within both categories are best conceptualized as neurobiological phenomena organized within distributed neural networks of the brain.

This portion of the mental status examination is crucial not only for recognizing psychiatric disorders that may come to the attention of the behavioral neurologist but also as a reminder that emotion and personality are subserved by the brain and can be interpreted in neurobehavioral terms (Cummings and Mega 2003). Mood and affect are key areas to assess, and information on changes in comportment may also emerge during this process. Although some view these aspects of the clinical evaluation as more properly assigned to psychiatrists or psychologists, mood, affect, and comportment have important neurobehavioral implications, and disturbances may prove critical in establishing a diagnosis. Assessment of these clinical features is necessarily subjective, and considerable experience is often needed to detect subtle alterations.

Mood is considered the content of an individual's emotional experience and is inaccessible to the examiner except through the patient's own statements (Cummings and Mega 2003). Affect, on the other hand, is a more observable phenomenon, consisting of the outward manifestations of emotions—facial expressions, tone of voice, latency of response, and paralinguistic communication such as gesture and body posture (Cummings and Mega 2003). In pursuing mood, the patient's own comments are often verified or contradicted by observations of family or friends, and interviews with these outside sources are often quite helpful. Detection of the vegetative signs of depression—weight loss, insomnia, anhedonia—is obviously critical at this point. Systematic questioning using the mnemonic SIGECAPS is helpful for gathering more detail; these questions probe sleep, interest, guilt, energy, concentration, appetite, psychomotor function, and suicidality (Weissberg 2004). The last of these queries is particularly essential, as the potential for self-harm or suicide, if detected, must provoke immediate preventive action. Affect is most often congruent with mood, so that a patient with a sad affect usually does feel depressed, whereas an elated, euphoric patient usually feels manic. One must bear in mind, however, that affect may not accurately reflect mood, particularly in patients with cerebral disease. The pathologic crying or laughter of patients with *pseudobulbar affect* from bifrontal disease, for example, is often incongruent with their mood, and patients with right hemisphere lesions may appear indifferent but actually be depressed (Cummings and Mega 2003). A final point is that primary motor disorders must not be mistaken for an affective disturbance; the masked facies of Parkinson's disease, for example, is not indicative of flat affect.

Finally, the area of *comportment* deserves attention since it may be the most helpful sign of a change in personality (Mesulam 2000). Personality is normally established in childhood and adolescence, and a change in personality in adulthood portends a neurologic disorder, usually of the frontal or temporal lobes. How a patient comports him- or herself may reveal a great deal about how the brain is working. An impairment in the integration of emotional behavior may be obvious from the history, and patients may also display such behaviors as sexual inappropriateness, apathy, abulia, facetiousness, and irritability—all commonly associated with bifrontal or frontal systems pathology. A particularly prominent feature of many patients with frontal lobe disorders is *disinhibition*, an often dramatic disorder of comportment that represents eroded control of limbic impulses and can be one of the most disabling neurobehavioral disturbances (Chapter 10). Disturbances of this kind are more difficult to measure and quantitate than problems in other neurobehavioral domains, such as aphasia, amnesia, and executive dysfunction, but may be obvious to all concerned or detectable with focused history-taking; in many cases, patients with disinhibition come to the attention of the legal system long before any physician is involved.

STANDARDIZED MENTAL STATUS TESTING

The procedures for evaluating mental status described above offer a general framework to help organize the practical testing of clinically relevant behavior. This kind of evaluation allows the clinician to be comprehensive enough to sample all major neurobehavioral domains yet flexible enough to adapt the process to focus on salient problems that may appear. The result is a clinical impression that does not produce a single score for a patient's performance but rather a richly descriptive portrayal of deficits and strengths that captures the essence of the individual's neurobehavioral profile.

At times, however, this somewhat subjective approach can be complemented by more objective evaluation, and the use of standardized tests that generate comparative scores is often helpful (Weintraub 2000; Lezak et al. 2004). One popular such test is the Mini-Mental State Examination (MMSE; Folstein, Folstein, and McHugh 1975), which yields a score from 0 to a maximum of 30 and requires only 5 to 10 minutes for administration (Table 2.2). Knowledge of the MMSE and its utility is important because it is widely used by practitioners

TABLE 2.2. Mini–Mental State Examination

Maximum Score	Score	
		Orientation
5	( )	What is the (year) (season) (date) (day) (month)?
5	( )	Where are we: (state) (county) (town) (hospital) (floor)?
		Registration
3	( )	Name three objects, one second to say each, then ask the patient to repeat all three after you have said them. Give one point for each correct answer. Continue repeating all three objects until the patient learns all three.
		Attention and Calculation
5	( )	Serial sevens. One point for each correct response. Stop after five answers. Alternatively, spell "world" backwards.
		Recall
3	( )	Ask for the three objects named in *Registration*. Give one point for each correct answer.
		Language
2	( )	Name a pencil and watch.
1	( )	Repeat the following: "No ifs, ands, or buts."
3	( )	Follow a three-stage command: "Take this paper in your right hand, fold it in half, and put it on the floor."
1	( )	Read and obey the following: "CLOSE YOUR EYES."
1	( )	Write a sentence.
1	( )	Copy a design.
30		**Total**

Source: M. F. Folstein, S. E. Folstein, and P. R. McHugh. 1975. "Mini-mental state." A practical method for grading the cognitive state of patients for the clinician. *J. Psychiatr Res;* 12: 189–198. Reprinted with permission from Elsevier Science, Ltd., Pergamon Imprint, Oxford, England.

in many settings who seek to screen patients for cognitive disorders, and it is virtually ubiquitous as an outcome measure in clinical trials of drugs for cognitive disorders such as AD.

Despite the advantages of its brevity and the fact that patients with neurologic diseases usually score lower than those with psychiatric illnesses (Lezak et al. 2004), the MMSE is not a substitute for careful mental status evaluation as described above because it is too limited. For example, the MMSE is heavily weighted to assess language and memory, allocating only one point for visuo-spatial skills, so that a patient with a large right hemisphere lesion might well be misdiagnosed as normal because of a near-perfect score of 29/30. The MMSE and similar tests, however, are useful in the longitudinal evaluation of dementia patients; in general, patients with AD can be expected to decline by three points per year on the MMSE (Salmon et al. 1990).

A useful screening test for frontal systems impairment is the Frontal Assessment Battery (FAB; Dubois et al. 2000), which assesses executive function and related domains. The FAB is convenient to administer and generates a score ranging from 0 to 18, with scores below 16 considered abnormal. This test is sensitive to a wide range of frontal lobe disorders, including not only structural lesions but also metabolic disturbances and psychiatric conditions such as depression that can disrupt frontal systems functions. More recently, the Montreal Cognitive Assessment (MoCA) has been introduced and shows promise for combining aspects of the MMSE and FAB into a single 30-point measure (Nasreddine et al. 2005).

More detailed standardized testing, of course, is available by referral to a neuropsychologist (Weintraub 2000; Lezak et al. 2004). Neuropsychology is a branch of psychology that is derived in large part from the work of the Russian neuropsychologist A. R. Luria, who in the mid-twentieth century pioneered the systematic assessment of altered behavior in Russian war veterans (Luria 1980). Neuropsychological testing extends and elaborates the mental status evaluation, providing a broad range of objective data and concluding with a summary opinion of an individual's cognitive and emotional status. The information gathered by neuropsychological assessment is particularly helpful if it is unclear whether the patient has neurologic or psychiatric illness and in establishing the pattern and severity of deficits in patients with known structural damage. These data may therefore prove highly valuable in diagnosis, treatment, rehabilitation, counseling, and litigation issues (Lezak et al. 2004).

While it is not within our scope to discuss neuropsychology in detail, a brief comment on its methods is helpful. Testing usually requires 4 to 8 hours, and a full day is devoted to the evaluation so as to ensure adequate break time and minimize the potential stress of testing. A wide variety of tests can be administered to probe aspects of cognitive function and emotional status. Two broad philosophies are represented in the field, one adhering to highly structured test batteries such as the Halstead-Reitan Battery (Russell, Neuringer, and Goldstein 1970) and the Luria-Nebraska Battery (Golden 1981) from which a myriad of scores are generated, and the other advocating a "process approach" in which the unique manner in which a patient performs a task is given particular attention (Kaplan 1983). In the former, a quantitative evaluation is considered paramount, whereas the latter stresses qualitative assessment as most revealing. Neuropsychologists

TABLE **2.3.** Popular neuropsychological tests

General Intellectual Ability
Wechsler Adult Intelligence Scale-Four (WAIS-4; Wechsler, Coalson, and Raiford 2008)
Dementia Rating Scale (DRS; Mattis 1988)
Mini-Mental State Examination (MMSE; Folstein et al. 1975)
Repeatable Battery for Assessment of Neuropsychological Status (RBANS; Randolph et al. 1998)
The Montreal Cognitive Assessment (MoCA; Nasreddine et al. 2005)

Attention
Digit Span (Cummings and Mega 2003)
Digit Vigilance Test (Lewis and Rennick 1979)
Trail Making Test (Reitan and Davison 1974)
Paced Auditory Serial Addition Test (PASAT; Gronwall 1977)

Memory
Wechsler Memory Scale-Four (Wechsler, Holdnack, and Drozdick 2009)
California Verbal Learning Test (CVLT; Delis et al. 1987)
Rey Auditory Verbal Learning Test (Lezak et al. 2004)

Language
Boston Diagnostic Aphasia Examination (BDAE; Goodglass and Kaplan 1983)
Boston Naming Test (Kaplan, Goodglass, and Weintraub 1983)
Verbal Associative Fluency Test (Benton 1968)
Aphasia Screening Test (Reitan 1984)

Visuospatial Ability
Rey-Osterrieth Complex Figure Test (Lezak et al. 2004)
Boston Parietal Drawings (Goodglass and Kaplan 1983)
Hooper Visual Organization Test (Hooper 1958)
Judgment of Line Orientation Test (Benton, Varney, and Hamsher 1978)

Executive Function
Wisconsin Cart Sorting Test (WCST; Grant and Berg 1948)
Category Test (Reitan and Davison 1974)
Stroop Test (Stroop 1935)
Raven's Progressive Matrices (Raven 1960)
Frontal Assessment Battery (FAB; Dubois et al. 2000)

Emotion and Personality
Beck Depression Inventory (Beck et al. 1961)
Minnesota Multiphasic Personality Inventory–2 (MMPI-2; Butcher 2000)
Behavioral Dyscontrol Scale (Grigsby, Kaye, and Robbins 1992)

---

usually adopt an intermediate position, combining scores that allow comparison of a patient with many others who have been similarly tested and individualized assessments of performance to characterize particular neuropsychological features (Weintraub 2000; Lezak et al. 2004).

An impressive number of tests have been assembled by neuropsychologists to probe the various areas of neurobehavioral function (Weintraub 2000; Lezak et al. 2004). Table 2.3 lists some of the more commonly used neuropsychological measures, grouped by the major functional domain they assess. This list is far from exhaustive and emphasizes measures that correspond reasonably well to the cognitive and emotional domains considered in this book. Many other neuropsychological tests may yield useful information depending on the patient under consideration (Lezak et al. 2004).

REFERENCES

Adolphs, R. The social brain: neural basis of social knowledge. *Annu Rev Psychol* 2009; 60: 693–716.

Alexander, M. P., Stuss, D. T., and Benson, D. F. Capgras syndrome: a reduplicative phenomenon. *Neurology* 1979; 29: 334–339.

American Psychiatric Association. *Diagnostic and Statistical Manual of Mental Disorders*. 4th ed. Washington, DC: American Psychiatric Association; 1994.

Anderson, C. A., Camp, J., and Filley, C. M. Erotomania after aneurysmal subarachnoid hemorrhage: case report and literature review. *J Neuropsychiatry Clin Neurosci* 1998; 10: 330–337.

Austin, J. H. *Chase, Chance, and Creativity*. 2nd ed. Cambridge: MIT Press; 2003.

Baddeley, A. Working memory. *Science* 1992; 255: 556–559.

Beck, A. T., Ward, C. H., Mendelson, M., et al. An inventory for measuring depression. *Arch Gen Psychiat* 1961; 4: 561–571.

Benson, D. F., Gardner, H., and Meadows, J. C. Reduplicative paramnesia. *Neurology* 1976; 26: 147–151.

Benton, A. L. Differential behavioral effects in frontal lobe disease. *Neuropsychologia* 1968; 6: 53–60.

Benton, A. L., Varney, N. R., and Hamsher, K. de S. Visuospatial judgment: a clinical test. *Arch Neurol* 1978; 35: 364–367.

Berrios, G. E. Tactile hallucinations: conceptual and historical aspects. *J Neurol Neurosurg Psychiatry* 1982; 45: 285–293.

Butcher, J. N. *Basic Sources on the MMPI-2*. Minneapolis: University of Minnesota Press; 2000.

Cummings, J. L. Organic delusions: phenomenology, anatomical correlations, and review. *Br J Psychiat* 1985; 146: 184–197.

———. Psychosis in neurologic disease. *Neuropsychiatry Neuropsychol Behav Neurol* 1992; 5: 144–150.

Cummings, J. L., and Mega, M. S. *Neuropsychiatry and Behavioral Neuroscience*. New York: Oxford University Press; 2003.

Daly, D. D. Ictal clinical manifestations of complex partial seizures. *Adv Neurol* 1975; 11: 57–82.

Damas-Mora, J., Skelton-Robinson, M., and Jenner, F. A. The Charles Bonnet syndrome in perspective. *Psychol Med* 1982; 12: 251–261.

Delis, D. C., Kramer, J. H., Kaplan, E., and Ober, B. A. *California Verbal Learning Test*. San Antonio: Psychological Corporation; 1987.

Dubois, B., Slachevsky, A., Litvan, I., and Pillon, B. The FAB: a frontal assessment battery at bedside. *Neurology* 2000; 55: 1621–1626.

Dunn, D. W., Weisberg, L. A., and Nadell, J. Peduncular hallucinations caused by brainstem compression. *Neurology* 1983; 33: 1360–1361.

Feinberg, T. E., Eaton, L. A., Roane, D. M., and Giacino, J. T. Multiple Fregoli delusions after traumatic brain injury. *Cortex* 1999; 35: 373–387.

Filley, C. M. The frontal lobes. In: Boller, F., Finger, S., and Tyler, K. L., eds. *Handbook of Clinical Neurology*. Edinburgh: Elsevier; 2009: 95: 557–570.

———. The neuroanatomy of attention. *Semin Speech Lang* 2002; 23: 89–98.

Filley, C. M., and Jarvis, P. E. Delayed reduplicative paramnesia. *Neurology* 1987; 37: 701–703.

Folstein, M. F., Folstein, S. E., and McHugh, P. R. "Mini-mental state": a practical method for grading the cognitive state of patients for the clinician. *J Psychiatr Res* 1975; 12: 189–198.

Frith, C. Social cognition. *Phil Trans R Soc B* 2008; 363: 2033–2039.

Golden, C. J. A standardized version of Luria's neuropsychological tests: a quantitative and qualitative approach to neuropsychological evaluation. In: Filskov, S. B., and Boll, T. J., eds. *Handbook of Clinical Neuropsychology*. New York: Wiley-Interscience; 1981: 608–642.

Goodglass, H., and Kaplan, E. *The Assessment of Aphasia and Related Disorders*. 2nd ed. Philadelphia: Lea and Febiger; 1983.

Grant, D. A., and Berg, E. A. A behavioral analysis of degree of reinforcement and ease of shifting to new responses in a Weigl-type cart-sorting problem. *J Exp Psychol* 1948; 38: 404–411.

Grigsby, J., Kaye, K., and Robbins, L. J. Reliabilities, norms and factor structure of the Behavioral Dyscontrol Scale. *Percept Mot Skills* 1992; 74: 883–892.

Gronwall, D.M.A. Paced auditory serial addition task: a measure of recovery from concussion. *Percep Motor Skills* 1977; 44: 367–373.

Hecaen, H., and Albert, M. L. *Human Neuropsychology*. New York: John Wiley and Sons; 1978.

Heilman, K. M. *Creativity and the Brain*. New York: Psychology Press; 2005.

Hillis, A. E. Aphasia: progress in the last quarter of a century. *Neurology* 2007; 69: 200–213.

Hooper, H. E. *The Hooper Visual Organization Test Manual*. Los Angeles: Western Psychological Services; 1958.

Jacobs, L., Feldman, M., Diamond, S. P., and Bender, M. B. Palinacousis: persistent or recurring auditory sensations. *Cortex* 1973; 9: 275–287.

James, W. *The Principles of Psychology*. New York: Henry Holt; 1890.

Kaplan, E. Process and achievement revisited. In: Wapner, S., and Kaplan, B., eds. *Toward a Holistic Developmental Psychology*. Hillsdale, NJ: Lawrence Erlbaum; 1983: 143–156.

Kaplan, E., Goodglass, H., and Weintraub, S. *Boston Naming Test*. Philadelphia: Lea and Febiger; 1983.

King, D. W., and Marsan, C. A. Clinical features and ictal patterns in epileptic patients with EEG temporal lobe foci. *Ann Neurol* 1977; 2: 138–147.

Lepore, F. E. Spontaneous visual phenomena with visual loss: 104 patients with lesions of retinal and neural afferent pathways. *Neurology* 1990; 40: 444–447.

Lessell, S. Higher disorders of visual function: positive phenomena. In: Glaser, J. S., and Smith, J. L., eds. *Neuro-ophthalmology*, vol. VIII. St Louis: C. V. Mosby; 1975: 27–44.

Lewis, R. F., and Rennick, P. M. *Manual for the Repeatable Cognitive-Perceptual-Motor Battery.* Grosse Pointe Park, MI: Axon Publishing Co.; 1979.

Lezak, M. D., Howieson, D. B., Loring, D. W., et al. *Neuropsychological Assessment.* 4th ed. New York: Oxford University Press; 2004.

Lipowski, Z. J. *Delirium: Acute Brain Failure in Man.* Springfield, IL: Charles C. Thomas; 1980.

Luria, A. R. *Higher Cortical Functions in Man.* New York: Consultants Bureau; 1980.

Mattis, S. *Dementia Rating Scale.* Odessa, FL: Psychological Assessment Resources; 1988.

Mesulam, M.-M. Behavioral neuroanatomy: large-scale networks, association cortex, frontal systems, the limbic system, and hemispheric specializations. In: Mesulam, M.-M. *Principles of Behavioral and Cognitive Neurology.* 2nd ed. New York: Oxford University Press; 2000: 1–120.

Miller, B. L., and Cummings, J. L., eds. *The Human Frontal Lobes: Functions and Disorders.* New York: Guilford Press, 1999.

Nasreddine, Z. S., Phillips, N. A., Bédirian, V., et al. The Montreal Cognitive Assessment (MoCA): a brief screening test for mild cognitive impairment. *J Am Geriatr Soc* 2005; 53: 695–699.

Posner, J. B., Saper, C. B., Schiff, N. D., and Plum, F. *Plum and Posner's Diagnosis of Stupor and Coma.* 4th ed. New York: Oxford University Press; 2007.

Randolph, C., Tierney, M. C., Mohr, M., and Chase, T. N. The Repeatable Battery for the Assessment of Neuropsychological Status (RBANS): preliminary clinical validity. *J Clin Exp Neuropsychol* 1998; 20: 310–319.

Reitan, R. M. *Aphasia and Sensory-Perceptual Deficits in Adults.* Tucson: Reitan Neuropsychology Laboratories; 1984.

Reitan, R. M., and Davison, L. A. *Clinical Neuropsychology: Current Status and Applications.* New York: Hemisphere; 1974.

Rilling, J. K., Gutman, D. A., Zeh, T. R., et al. A neural basis for social cooperation. *Neuron* 2002; 35: 395–405.

Ross, E. D. Affective prosody and the aprosodias. In: Mesulam, M.-M. *Principles of Behavioral and Cognitive Neurology.* 2nd ed. New York: Oxford University Press; 2000: 316–331.

Russell, E. W., Neuringer, C., and Goldstein, G. *Assessment of Brain Damage: A Neuropsychological Key Approach.* New York: Wiley Interscience; 1970.

Salmon, D. P., Thal, L. J., Butters, N., and Heindel, W. C. Longitudinal evaluation of dementia of the Alzheimer type: a comparison of 3 standardized mental status examinations. *Neurology* 1990; 40: 1225–1230.

Shepherd, M. Morbid jealousy: some clinical and social aspects of a psychiatric symptom. *J Ment Sci* 1961; 107: 687–753.

Siegel, R. K. Hallucinations. *Sci Am* 1977; 237: 132–140.

Stagno, S. J., and Gates, T. J. Palinopsia: a review of the literature. *Behav Neurol* 1991; 4: 67–74.

Stroop, J. R. Studies of interference in serial verbal reactions. *J Exp Psychol* 1935; 18: 643–662.

Strub, R. L., and Black, F. W. *The Mental Status Examination in Neurology.* 3rd ed. Philadelphia: F. A. Davis; 1993.

Stuss, D. T., and Benson, D. F. *The Frontal Lobes*. New York: Raven Press; 1986.

Tranel, D., Rudrauf, D., Vianna, E. P., and Damasio, H. Does the Clock Drawing Test have focal neuroanatomical correlates? *Neuropsychology* 2008; 22: 553–562.

Victor, M., and Hope, J. M. The phenomenon of auditory hallucinations in chronic alcoholism. *J Nerv Ment Dis* 1958; 126: 451–481.

Wechsler, D., Coalson, D. L., and Raiford, S. E. *Wechsler Adult Intelligence Test: Fourth Edition Technical and Interpretive Manual*. San Antonio: Pearson; 2008.

Wechsler, D., Holdnack, J. A., and Drozdick, L. W. *Wechsler Memory Scale: Fourth Edition Technical and Interpretive Manual*. San Antonio: Pearson; 2009.

Weintraub, S. Neuropsychological assessment of mental state. In: Mesulam, M.-M. *Principles of Cognitive and Behavioral Neurology*. 2nd ed. New York: Oxford University Press; 2000: 121–173.

Weissberg, M. *The Colorado Medical Student Log of Basic Psychiatry*. 12th ed. Denver: Department of Psychiatry, University of Colorado School of Medicine; 2004.

Zeki, S. The neurobiology of love. *FEBS Letters* 2007; 581: 2575–2579.

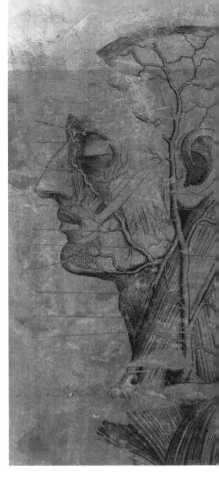

CHAPTER THREE

DISORDERS OF AROUSAL
AND ATTENTION

Disturbances of arousal and attention compose a heterogeneous and challenging group of neurobehavioral syndromes (Geschwind 1982; Mesulam 2000; Posner et al. 2007). Ranging in severity from coma after brainstem infarction to subtle acute confusional states related to drug intoxication, these disorders are not only common clinically but provide many insights into the brain's capacity to enable uniquely human mental life. Moreover, they bear directly upon the fundamental question of the nature of consciousness, a perennial philosophical conundrum that can now be meaningfully addressed in a neuroscientific context. This chapter will jointly consider the arousal and attentional disorders in some detail, as they are closely linked in the mental status examination and have several clinical and neuroanatomic similarities. First, however, some general background related to these topics merits consideration.

For any higher mental function to occur, the human brain must possess a mechanism that can maintain the waking state and another that permits the ability to focus awareness on behaviorally relevant external and internal stimuli. Both of these systems are indeed present, and it is useful and widely accepted to refer to them respectively as the arousal system and the attentional system. In everyday terms, the simple observation that one can be awake without being attentive suggests that arousal and attention are indeed separable. In a neurologic context, individuals who are in a vegetative state provide a dramatic clinical example of this distinction: in these unfortunate individuals, there is massive bilateral cerebral hemispheric damage, with sparing of the brainstem, leaving a patient with intact arousal but absent attention (Multi-Society Task Force on PVS 1994).

Arousal refers to the phenomenon of wakefulness, in most cases a readily observable state (Posner et al. 2007). Disorders of arousal imply a departure from normal wakefulness, excluding, of course, the physiologic process of sleep (Hobson 2005). *Attention* is a multidimensional concept with many meanings (Mesulam 2000), but its most familiar form is selective attention, the capacity to direct awareness to specific aspects of the extrapersonal space and simultaneously screen out competing but irrelevant information. When selective attention is operative over an extended period, sustained attention is said to be active, and alternative terms for this phenomenon are vigilance and concentration. Another pertinent concept is directed attention, which refers to selective attention directed to the contralateral hemispace.

These disorders all imply some form of dysfunction in the realm of consciousness, which has been the subject of renewed neuroscientific interest for several decades. Neurologists have found it useful to distinguish between the *level* of consciousness, or the degree of arousal, and the *content* of consciousness, the sum of all mental functions (Posner et al. 2007); attention as reviewed above would thus be one aspect of the content of consciousness. Arousal can be envisioned figuratively as the foundation of consciousness that permits wakefulness; this requirement met, the full expression of consciousness then involves the participation of all cognitive and emotional domains. As will be seen, disorders of the level and content of consciousness can be clinically dissociated, and each has a relatively secure, although often overlapping, neuroanatomy. Coma, the prototype syndrome of impaired arousal, results most directly from discrete lesions of the brainstem and thalamus, whereas acute confusional state, the classic syndrome of impaired attention, generally follows hemispheric dysfunction. It follows that a disorder of arousal always involves an attentional disorder as well, but the converse is not true; an acute confusional state may affect attention while largely sparing arousal.

Implicit in this discussion, of course, is the notion of consciousness itself. Many would concur with the position of the neurologic community that takes consciousness to mean the awareness of self and the environment (Posner et al. 2007). Yet despite this sturdy conception, and the convenient classifications of impaired consciousness that clinicians and researchers have put to good use, there is still something unfathomable about this most important property of the human mind. There is, to be sure, no reason to doubt that the structure and function of the brain account in some way for the phenomenon of consciousness, but the actual means by which neural activity is transformed into subjective conscious experience remains as yet an impenetrable mystery.

AROUSAL DYSFUNCTION

The anatomy of arousal has been elucidated in considerable detail, although uncertainties remain. Classic experiments using laboratory animals in the mid-twentieth century (Moruzzi and Magoun 1949) convincingly established that arousal is dependent on a widely distributed structure in the rostral brainstem that has come to be called the ascending reticular activating system (ARAS). Because it has indistinct boundaries throughout its extent, the ARAS is more a physiologic concept than a neuroanatomic entity; it can be visualized, however, as a portion of the reticular formation in the core of the brainstem extending from the pons to the thalamus via the midbrain (Figure 3.1). Although the ARAS receives important input from all sensory modalities—accounting for the rapid arousal that ensues when, for example, one touches a hot stove—there are endogenous mechanisms for arousal as well. One example of such an intrinsic mechanism is the cyclical alternation between wakefulness and sleep, which, while not completely understood, clearly depends in large measure on active neurochemical processes occurring in the ARAS and hypothalamus (Hobson 2005). The ARAS sends projections to the intralaminar nuclei of the thalamus—the two largest of which are the centromedian and parafascicular nuclei—and from these midline relay stations further connections are made to widespread areas of the cerebral cortex (Morison and Dempsey 1942; Figure 3.1). The role of the ARAS and thalamocortical projection system is graphically evident in the electrical activity obtained from electroencephalography, which records rhythmic oscillations on the scalp, reflecting the synchronous firing of neural systems subserving arousal and attention.

Clinical disorders of arousal may take a number of forms, resulting in either hyperaroused or, more commonly, hypoaroused states. Hyperarousal can appear as restlessness, agitation, or delirium, all presumably due to some loss of inhibitory control in the hemispheres that releases lower structures from normal regulation. Hypoarousal can take many forms, depending on the severity of the causative condition. Less severe deficits can be roughly described by use of a spectrum of descriptors ranging from drowsiness and somnolence to lethargy and obtundation. More dramatic are the two syndromes of *stupor*, a state of unresponsiveness from which arousal can be achieved only by vigorous and repeated stimulation, and *coma*, a state of unarousable unresponsiveness

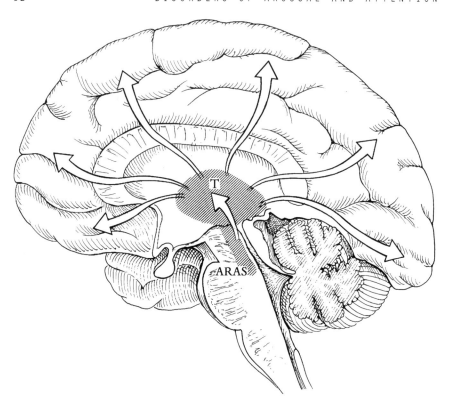

FIGURE 3.1. *Midsagittal view of the brain illustrating structures responsible for arousal: the ascending reticular activating system (ARAS) and the thalamus (T).*

(Posner et al. 2007). The clinical neurologic literature supports the thesis that the arousal system in the upper brainstem and thalamus does indeed maintain alertness (Brain 1958; Katz, Alexander, and Mandell 1987), and a variety of structural lesions in these regions, including infarcts, hemorrhages, tumors, and abscesses, have caused stupor and coma. As a general rule, small and restricted lesions in the arousal system can be sufficient to result in stupor or coma, whereas hemispheric lesions must be extensive and bilateral to produce the same picture (Posner et al. 2007). Deep midline lesions of the upper brainstem and diencephalon, therefore, are most clearly associated with disorders of arousal.

Other related conditions require brief consideration before leaving the topic of arousal dysfunction. These are all disorders in which the paucity or absence of limb movement and speech suggests an arousal disorder, but which in fact do not involve this kind of disturbance. The devastating syndrome of the *persistent vegetative state* (PVS) implies massive bilateral hemispheric insult from trauma, anoxia, degenerative disease, or other etiologies and deprives the patient of the content of consciousness while sparing arousal (Multi-Society Task Force on PVS 1994). Such a patient has preserved sleep-wake cycles and appears, during waking, to be alert but has no self-awareness or other evidence of higher function. *Akinetic mutism* is similar to PVS but involves less widespread destruction, often in medial frontal areas, and patients display an extreme form of abulia (Chapter 10). Arousal and sleep-wake cycles are normal, but there is absent motivation to perform any motor or mental action. Finally, the *locked-in syndrome* involves no mental deficit at all, but patients are de-efferented by a lesion in the pons that spares the ARAS but causes quadriplegia (Posner et al. 2007). Only by careful examination using vertical eye movements and eye blinking—preserved because they are mediated by the midbrain—will it be apparent that no alteration of consciousness is present. All three of these conditions need to be distinguished from stupor and coma.

Most recently, the *minimally conscious state* (MCS) has been introduced as another important alteration of consciousness relevant to the concept of arousal. After consensus meetings sponsored by the Aspen Neurobehavioral Conference in the 1990s, the syndrome of MCS was proposed to describe patients who have inconsistent but clearly discernible behavioral evidence of consciousness that distinguishes the syndrome from coma and PVS (Giacino et al. 2002). One of the motives fueling this development was the frequent observation that PVS is not always persistent, in that some patients make unexpected but genuine recovery after variable time periods have elapsed. Thus MCS was established as a concept to designate patients who evolve from coma or PVS to a higher level of cognition (Giacino et al. 2002). MCS can also appear in the course of degenerative or congenital nervous system disorders, in which case the syndrome is a more severe manifestation of a progressive disease process (Giacino et al. 2002). In MCS patients who show clinical improvement, it is not clear how much further progress can be gained in an individual patient, but in any case, the recognition of MCS as a syndrome has opened up a host of crucial clinical issues related to prognosis, treatment, and rehabilitation of individuals who were formerly thought to have a dismal outcome. The widespread acceptance of this new syndrome (Posner et al. 2007) indicates a gratifying development in the sophistication with which these complex clinical problems are approached, and MCS has refined our understanding of consciousness and its disorders as well as generated a host of new clinical research opportunities.

The arousal system deep in the brain is clearly necessary for human mental life, but is far from sufficient. The hemispheres, in contrast, are dedicated to the content of consciousness: the totality of all the human cognitive and emotional functions. The remainder of this book deals with disorders in which the content of consciousness is specifically affected.

ATTENTIONAL DYSFUNCTION
....

Attention, a complex psychological concept, is more difficult to relate to brain structure than arousal. Since attentional competence underlies all higher function, attention plays a role in every conscious action and is therefore widely distributed across the entire brain. Still, clinical observations and recent functional imaging data indicate that certain regions play a special role in this fundamental neurobehavioral domain. We have seen how brainstem and thalamic structures regulate arousal, and we now proceed to the cerebral hemispheres as areas more involved in attention. The common neurobehavioral syndrome known as the acute confusional state will serve as a vivid illustration of the clinical importance and neuroanatomic basis of disordered attention.

One of the most common conditions encountered in medical and neurologic practice is the *acute confusional state* (Strub 1982; Taylor and Lewis 1993; Cummings and Mega 2003). Often the term "altered mental state" is used to describe this syndrome, but suffers from being nonspecific since most disorders discussed in this book could be so described. The abrupt appearance of confusion in a previously healthy person signifies a serious disturbance of the brain that can have one or more of many etiologies. Recognition of the acute confusional state is crucial because most of its causes are reversible if promptly treated (Strub 1982; Taylor and Lewis 1993; Cummings and Mega 2003). The reversibility of this syndrome is due to the fact that most cases involve toxic or metabolic disorders that exert a widespread but not initially destructive effect upon the brain. Only if the insult persists without treatment for extended periods does the possibility of irreversible structural damage arise. In the less common cases where structural damage is the immediate cause of the syndrome, the outcome may be less favorable.

Confusion is a problematic word since it has such a strong ordinary meaning that interferes with its medical specificity. In neurobehavioral terms, *confusion* is best thought of as the inability to maintain a coherent line of thought (Geschwind 1982). Confused patients have difficulty attending to meaningful environmental stimuli, while at the same time distractibility results from undue attention to irrelevant ones. The acute confusional state refers to a rapidly evolving disorder of selective attention that precludes the adequate performance of other neurobehavioral operations in the domains of memory, language, praxis, visuospatial skills, gnosis, and executive function (Chedru and Geschwind 1972a). To a lesser extent, arousal is also affected, and, in practice, arousal and attentional deficits are often commingled. Some cases, in fact, evolve from arousal deficits acutely to persistent attentional disturbances (Katz, Alexander, and Mandell 1987).

Alternate designations for the acute confusional state are commonly used and depend to a large extent on the clinical context. Some authorities, usually from the perspective of internal medicine, use the term *toxic-metabolic encephalopathy*, which has the advantage of indicating the most common causes of the syndrome and the fact that they affect the brain diffusely. Others follow the psychiatric custom and prefer the colorful word *delirium*, implying a state of agitated confusion with associated autonomic overactivity (American Psychiatric Association 1994). In the context of behavioral neurology, however, acute confusional state is preferred because it identifies the fundamental clinical impairment that is invariably present: the disorder of attention known as confusion. Furthermore, toxic-metabolic encephalopathy fails to include other causes of the syndrome, such as trauma, seizures, and stroke, and delirium neglects the hypoactive or lethargic confusional states without autonomic overactivity that are in fact more common than the hyperactive and agitated states of delirium (Lipowski 1990). The term "acute organic brain syndrome" is too vague to be of any use and is not recommended.

Clinical features of the acute confusional state are usually readily apparent. *Inattention* with poor concentration, vigilance, and mental control is the central deficit, and *disorientation* and *distractibility* are also typically present. Performance on the digit span test is reduced, reflecting an impairment in the brain's normal capacity to process information (Miller 1956). A characteristic finding is a fluctuating level of consciousness, which is not ordinarily seen in dementia, aphasia, and psychiatric disorders, conditions often mistaken for acute confusional state (Cummings and Mega 2003). Other prominent manifestations that may appear are bizarre but thematically consistent naming errors known as *nonaphasic misnaming* (Weinstein and Keller 1963) and disordered writing with spatial aberrations and spelling errors (Chedru and Geschwind 1972b). Memory deficits, insofar as they can be evaluated, are present. Aphasia in the usual sense (Chapter 5) is not seen, although mild anomia, verbal paraphasias, and dysarthria are common. Visuospatial dysfunction is usually encountered. In *delirium tremens*, the most dramatic of the acute confusional states, there is profound agitation associated with hallucinations, delusions, insomnia, tachycardia, fever, and diaphoresis (Charness, Simon, and Greenberg 1989). Patients with acute confusional states can also manifest other neurologic signs such as action tremor, asterixis, myoclonus, increased tone, hyperreflexia, and extensor plantar responses. The syndrome is particularly common in the elderly, and manifestations in the evening hours ("sundowning") may be especially troublesome. Individuals with preexisting neurobehavioral impairment are also more prone to develop the acute confusional state (Lipowski 1990). The electroencephalogram is useful in diagnosis because of the typical appearance of generalized theta (4–7 Hz) or delta (1–3 Hz) range slowing (Obrecht, Okhomina, and Scott 1979), and in occasional cases the tracing discloses nonconvulsive status epilepticus, either complex partial or absence in origin, that explains the confusional state.

The causes of acute confusional state constitute a long list (Table 3.1). A great variety of intoxicants, metabolic disorders, infections, inflammatory diseases, traumatic injuries, vascular events, epileptic states, and neoplasms can prove to be responsible, and the postsurgical period is a frequent setting for its appearance (Strub 1982; Cummings and Mega 2003). The vascular syndromes are less well recognized but important: infarction in the distribution of the right middle cerebral artery (Mesulam et al. 1976), the anterior communicating artery (Alexander and Freedman 1984), and the posterior cerebral arteries (Medina, Rubino, and Ross 1974) may all cause an acute confusional state. In the best known example of a focal lesion producing this syndrome, destruction of the right parietal lobe is a plausible etiology because of the important role of this region in directed attention (Mesulam 1990, 2000). Psychiatric disorders such as schizophrenia or mania may present with features that mimic but are distinct from those that characterize the acute confusional state, and a patient with acute Wernicke's aphasia can be mistakenly diagnosed as acutely confused. Mental status evaluation proves invaluable in distinguishing among these possibilities, and the appropriate use of laboratory tests and neuroimaging is also important.

Treatment of the acute confusional state depends on its cause, which must be sought by a prompt and comprehensive neurologic and medical evaluation. While the cause is being sought and addressed, supportive treatment is also important (Lipowski 1987). In most cases the syndrome will be partially or completely reversible; often the simple withdrawal of centrally acting medications—especially in the elderly—results in a gratifying recovery. It is well to bear in mind, however, that recovery may require days to weeks in the elderly, particularly if more than one causative factor has been involved. Finally, it should be added that the acute confusional state can be fatal if the underlying conditions are sufficiently serious (Liston 1982).

Another syndrome that appears to involve attentional dysfunction is the controversial disorder known as *attention-deficit/hyperactivity disorder* (ADHD). Originally described in children of school age, ADHD has also been postulated to occur in adulthood as a "residual type" (Denckla 1991). This syndrome essentially refers to individuals who have difficulty paying attention, completing tasks, organizing activity, and planning ahead; many are hyperactive, impulsive, and restless, leading to a formal distinction between "inattentive" and "hyperactive-impulsive" subtypes (American Psychiatric Association 1994). Although there are as yet no demonstrable brain abnormalities in these individuals, recent findings have implicated dysfunction of both frontal lobes and right hemisphere systems. Neuropsychological evidence in ADHD children has suggested frontal but not temporal lobe dysfunction (Shue and Douglas 1992). In addition, functional imaging studies have found bilateral premotor and prefrontal hypometabolism in adults with hyperactivity of childhood onset (Zametkin et al. 1990). In con-

TABLE 3.1. Major etiologies of the acute confusional state

Toxic	Vascular
Prescription drugs	Stroke
Nonprescription drugs	Subarachnoid hemorrhage
Alcohol withdrawal	Subdural hematoma
Benzodiazepine withdrawal	Hypertensive encephalopathy
	Migraine
Metabolic	**Epileptic**
Hypoxia	Ictal state
Hypoglycemia	Postictal state
Uremia	Complex partial status epilepticus
Hepatic disease	Absence status epilepticus
Thiamine deficiency	
Electrolyte disturbances	
Endocrinopathies	
Infectious and Inflammatory	**Neoplastic**
Meningitis	Tumors
Encephalitis	Cerebral edema
Vasculitis	Increased intracranial pressure
Abscess	
Traumatic	**Postsurgical**
Concussion	Preoperative atropine
Moderate traumatic brain injury	Hypoxia
Severe traumatic brain injury	Analgesics
	Electrolyte imbalance
	Fever

trast, children with right hemisphere lesions have been found to have a high incidence of attention-deficit disorder (Voeller 1986). Moreover, findings of hemineglect in children with attention deficit disorder have suggested right hemisphere dysfunction (Voeller and Heilman 1988). By analogy with studies in adults showing a tendency of right frontal lesions to cause motor impersistence (Kertesz et al. 1985), it may be that ADHD represents a functional disturbance of the right frontal lobe. More recent work, however, has documented more widespread changes in the brain. In a large MRI volumetric study that included ADHD patients who had never been medicated, overall brain volume—including the gray and white matter of the cerebral hemispheres, the caudate, and the cerebellum—was reduced by about 3 percent; of note, this effect was most apparent in the cerebral white matter (Castellanos et al. 2002).

Although much progress has been made, a firm anatomy for attention remains to be established. As mentioned above, attention is required for all higher functions and thus can be said to be associated with all cerebral areas. However, certain regions do appear to be particularly affiliated with attention, and two parallel systems are most crucial for clinical purposes. A substantial literature exists on the prominent role of the right hemisphere in directed attention, and Chapter 8 will take up this topic in more detail. There is also good reason to believe that the frontal lobes and their connections to posterior regions mediate the phenomena of selective and sustained attention. Recent functional magnetic resonance imaging data, for example, identify an attentional network involving large frontoparietal regions bilaterally (Fan et al. 2005). We will consider in particular the frontal contribution to attention, as these lobes have the most relevance in terms of the global attentional disorders that are so common in practice. Figure 3.2 schematically illustrates the major structures involved in attention and an overall conception of the relationship of arousal to attention.

The frontal lobes appear to be particularly crucial to many aspects of attentional function (Luria 1973; Stuss and Benson 1984; Mesulam 2000; Stuss and Alexander 2007). Receiving extensive projections from the brainstem and thalamus, the frontal lobes send dense projections to posterior cortical and subcortical regions, most prominently the parietal lobes, and receive reciprocal projections in return. It is clear that the frontal lobes are richly connected with all other regions of the hemispheres and are thus anatomically situated to exert a regulatory and organizing influence on the rest of the brain (Chapter 10). Numerous case studies from the clinical literature amply verify that frontal lesions can have a profound effect on the attentional abilities of affected patients (Stuss and Benson 1984; Mesulam 2000; Stuss and Alexander 2007).

The prefrontal areas may be of particular importance with regard to attention (Weinberger 1993; Rossi et al. 2009). In keeping with the executive role of the dorsolateral frontal regions (Chapter 10), the prefrontal cortices are intimately involved with the mediation of attention and the performance of purposeful activity (Weinberger 1993; Rossi et al. 2009). Functional neuroimaging studies have found that normal individuals responding to incoming stimuli or performing mental operations activate prefrontal areas regardless of what posterior sensory or cognitive region is also activated. To illustrate, studies have shown that the prefrontal cortex is strongly activated whether subjects attend to visual, auditory, or tactile stimuli (Roland 1982), and that prefrontal regions are activated by both verbal and nonverbal auditory stimuli (Mazziotta et al. 1982).

Although both frontal lobes clearly play a major role in both selective and sustained attention, the right frontal lobe may be particularly competent. This

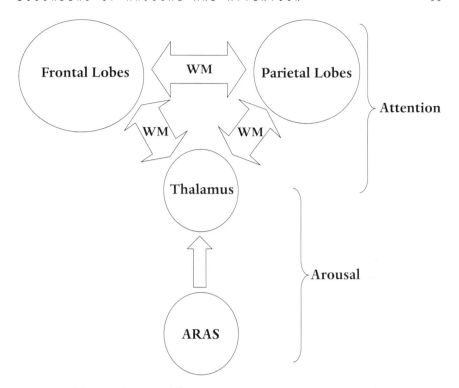

FIGURE 3.2. *Schematic depiction of the major structures involved in attention and a general conception of the neuroanatomic relationship between arousal and attention (ARAS: ascending reticular activating system; WM: white matter).*

advantage has been shown to be most apparent in terms of sustained attention or vigilance. Motor impersistence, as we have seen, is most typically seen after right frontal lesions (Kertesz et al. 1985). Neuropsychological evidence suggests that right frontal lesions are particularly detrimental to sustained attention (Wilkins, Shallice, and McCarthy 1987). Positron emission tomography (PET) data from normal subjects engaged in vigilance tasks (Cohen et al. 1988) and tasks requiring heightened effort (Bench et al. 1993) also implicate the right frontal lobe. Another PET study found both the right prefrontal and parietal areas to be active under similar conditions (Pardo, Fox, and Raichle 1991), leading to the idea that a network of right hemisphere structures is uniquely designed for the maintenance of sustained attention. Although it is unclear precisely how the frontal lobes and right hemisphere interact in the phenomena of attention, the right frontal lobe may represent the anatomic intersection of the distributed networks subserving selective attention.

Finally, there is likely to be a role of the cerebral white matter in the regulation of attention (Figure 3.2; Filley 1998, 2001). Because of the extensive white matter connections between the frontal lobes and the remainder of the cerebrum, most notably the parietal lobes, myelinated axons in the hemispheres play an important part in the regulatory functions of the frontal lobes that enable integrated cognitive and emotional operations. It is noteworthy that cortical deficits such as aphasia, apraxia, and agnosia are not typical of acute confusional states (Chedru and Geschwind 1972a), implying that dysfunction in structures underlying the neocortex is more implicated in the genesis of sustained attention deficits. A wide variety of cerebral white matter disorders have a prominent impact on sustained attention, leading to the idea that white matter plays a crucial role in the cerebral network mediating this capacity (Filley 1998, 2001). Reduced size of the corpus callosum, for example, has been linked with vigilance deficits in adults with multiple sclerosis (Rao et al. 1989) and has also been found in ADHD (Giedd et al. 1994), implying that frontal interhemispheric communication may be critical in the maintenance of sustained attention. We will consider the importance of the cerebral white matter in this and other areas of behavioral neurology more thoroughly in Chapter 12.

This discussion should not be interpreted as suggesting that the frontal lobes are dedicated centers for attention, or even that attentional disorders are reliably encountered after any frontal lobe lesion. Attention is a widely distributed phenomenon (Posner and Dehaene 1994; Mesulam 2000; Fan et al. 2005), and, as we have seen, it can be disrupted by a wide variety of lesions. Instead, the intent has been to illustrate how selective attention can be theoretically linked to the frontal regions with the aid of increasingly sophisticated clinical and experimental techniques. Although clearly not reducible to one cerebral area, attention appears to have a special affiliation with the frontal lobes, in keeping with the powerful organizational influence of frontal systems.

REFERENCES

Alexander, M. P., and Freedman, M. Amnesia after anterior communicating artery aneurysm rupture. *Neurology* 1984; 34: 752–757.

American Psychiatric Association. *Diagnostic and Statistical Manual of Mental Disorders.* 4th ed. Washington, DC: American Psychiatric Association; 1994.

Bench, C. J., Frith, C. D., Grasby, P. M., et al. Investigation of the functional anatomy of attention using the Stroop test. *Neuropsychologia* 1993; 31: 907–922.

Brain, R. The physiological basis of consciousness. *Brain* 1958; 81: 426–455.

Castellanos, F. X., Lee, P. P., Sharp, W., et al. Developmental trajectories of brain volume abnormalities in children and adolescents with attention-deficit/hyperactivity disorder. *J Am Med Assoc* 2002; 288: 1740–1748.

Charness, M. E., Simon, R. P., and Greenberg, D. A. Ethanol and the nervous system. *N Engl J Med* 1989; 321: 442–454.

Chedru, F., and Geschwind, N. Disorders of higher cortical functions in acute confusional states. *Cortex* 1972a; 8: 395–411.

———. Writing disturbances in acute confusional states. *Neuropsychologia* 1972b; 10: 343–354.

Cohen, R. M., Semple, W. E., Gross, M., et al. Functional localization of sustained attention: comparison to sensory stimulation in the absence of instruction. *Neuropsychiatry Neuropsychol Behav Neurol* 1988; 1: 3–20.

Cummings, J. L., and Mega, M. S. *Neuropsychiatry and Behavioral Neuroscience.* New York: Oxford University Press; 2003.

Denckla, M. B. Attention deficit hyperactivity disorder—residual type. *J Child Neurol* 1991; 6: S44–S48.

Fan, J., McCandliss, B. D., Fossella, J., et al. The activation of attentional networks. *Neuroimage* 2005; 26: 471–479.

Filley, C. M. The behavioral neurology of cerebral white matter. *Neurology* 1998; 50: 1535–1540.

———. *The Behavioral Neurology of White Matter.* New York: Oxford University Press; 2001.

Geschwind, N. Disorders of attention: a frontier in neuropsychology. *Phil Trans R Soc Lond* 1982; 298: 173–185.

Giacino, J. T., Ashwal, S., Childs, N., et al. The minimally conscious state: definition and diagnostic criteria. *Neurology* 2002; 58: 349–353.

Giedd, J. N., Castellanos, F. X., Casey, B. J., et al. Quantitative morphology of the corpus callosum in attention deficit hyperactivity disorder. *Am J Psychiatry* 1994; 151: 665–669.

Hobson, J. A. Sleep is of the brain, by the brain, and for the brain. *Nature* 2005; 437: 1254–1256.

Katz, D. I., Alexander, M. P., and Mandell, A. M. Dementia following strokes in the mesencephalon and diencephalon. *Arch Neurol* 1987; 44: 1127–1133.

Kertesz, A., Nicholson, I., Cancilliere, A., et al. Motor impersistence: a right hemisphere syndrome. *Neurology* 1985; 35: 662–666.

Lipowski, Z. J. Delirium (acute confusional states). *J Am Med Assoc* 1987; 258: 1789–1792.

———. *Delirium: Acute Confusional States.* New York: Oxford University Press; 1990.

Liston, E. H. Delirium in the aged. *Psychiat Clin N Am* 1982; 5: 49–66.

Luria, A. R. *The Working Brain.* New York: Basic Books; 1973.

Mazziotta, J. C., Phelps, M. E., Carson, R. E., and Kuhl, D. E. Tomographic mapping of human cerebral metabolism: auditory stimulation. *Neurology* 1982; 32: 921–937.

Medina, J. L., Rubino, F. A., and Ross, E. Agitated delirium caused by infarction of the hippocampal formation and fusiform and lingual gyri: a case report. *Neurology* 1974; 24: 1181–1183.

Mesulam, M.-M. Attentional networks, confusional states, and neglect syndromes. In: Mesulam, M.-M. *Principles of Behavioral and Cognitive Neurology.* 2nd ed. New York: Oxford University Press; 2000: 174–256.

———. Large-scale neurocognitive networks and distributed processing for attention, language, and memory. *Ann Neurol* 1990; 28: 597–613.

Mesulam, M.-M., Waxman, S. G., Geschwind, N., and Sabin, T. D. Acute confusional states with right middle cerebral artery infarctions. *J Neurol Neurosurg Psychiatry* 1976; 39: 84–89.

Miller, G. A. The magical number seven, plus or minus two: some limits on our capacity for processing information. *Psychol Bull* 1956; 63: 81–97.

Morison, R. S., and Dempsey, E. W. A study of thalamo-cortical relations. *Am J Physiol* 1942; 135: 281–292.

Moruzzi, G., and Magoun, H. W. Brainstem reticular formation and activation of the EEG. *Electroencephalogr Clin Neurophysiol* 1949; 1: 455–473.

Multi-Society Task Force on PVS. Medical aspects of the persistent vegetative state. *N Engl J Med* 1994; 330: 1499–1508, 1572–1579.

Obrecht, R., Okhomina, F.O.A., and Scott, D. F. Value of EEG in acute confusional states. *J Neurol Neurosurg Psychiatry* 1979; 42: 75–77.

Pardo, J. V., Fox, P. T., and Raichle, M. E. Localization of a human system for sustained attention by positron emission tomography. *Nature* 1991; 349: 61–64.

Posner, J. B., Saper, C. B., Schiff, N. D., and Plum, F. *Plum and Posner's Diagnosis of Stupor and Coma.* 4th ed. Oxford: Oxford University Press; 2007.

Posner, M. I., and Dehaene, S. Attentional networks. *Trends Neurosci* 1994; 17: 75–79.

Rao, S. M., Leo, G. J., Haughton, V. M., et al. Correlation of magnetic resonance imaging with neuropsychological testing in multiple sclerosis. *Neurology* 1989; 39: 161–166.

Roland, P. E. Cortical regulation of selective attention in man: a regional cerebral blood flow study. *J Neurophysiol* 1982; 48: 1059–1078.

Rossi, A. F., Pessoa, L., Desimone, R., and Ungerleider, L. G. The prefrontal cortex and the executive control of attention. *Exp Brain Res* 2009; 192: 489–497.

Shue, K. L., and Douglas, V. I. Attention deficit hyperactivity disorder and the frontal lobe syndrome. *Brain Cogn* 1992; 20: 104–124.

Strub, R. L. Acute confusional state. In: Benson, D. F., and Blumer, D., eds. *Psychiatric Aspects of Neurologic Disease*, vol. 2. New York: Grune and Stratton; 1982: 1–23.

Stuss, D. T., and Alexander, M. P. Is there a dysexecutive syndrome? *Philos Trans R Soc Lond B Biol Sci* 2007; 362: 901–915.

Stuss, D. T., and Benson, D. F. Neuropsychological studies of the frontal lobes. *Psychol Bull* 1984; 95: 3–28.

Taylor, D., and Lewis, S. Delirium. *J Neurol Neurosurg Psychiatry* 1993; 56: 742–751.

Voeller, K.K.S. Right-hemisphere deficit syndrome in children. *Am J Psychiatry* 1986; 143: 1004–1009.

Voeller, K.K.S., and Heilman, K. M. Attention deficit disorder in children: a neglect syndrome? *Neurology* 1988; 38: 806–808.

Weinberger, D. R. A connectionist approach to the prefrontal cortex. *J Neuropsychiatry* 1993; 5: 241–253.

Weinstein, E. A., and Keller, N.J.A. Linguistic patterns of misnaming in brain injury. *Neuropsychologia* 1963; 1: 79–90.

Wilkins, A. J., Shallice, T., and McCarthy, R. Frontal lesions and sustained attention. *Neuropsychologia* 1987; 25: 359–365.

Zametkin, A. J., Nordahl, T. E., Gross, M., et al. Cerebral glucose metabolism in adults with hyperactivity of childhood onset. *N Engl J Med* 1990; 323: 1361–1366.

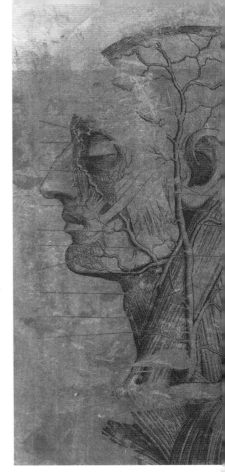

CHAPTER FOUR

MEMORY DISORDERS

Memory is a critical cognitive domain that has long challenged students of behavior. The importance of memory in human existence is indisputable, since without an intact mnemonic system, past experience cannot be called upon to deal with present contingencies or to formulate future plans. Many different disciplines have contributed to the study of memory, encompassing organisms as simple as the mollusk (Kandel 2006) and as complicated as higher primates (Zola-Morgan and Squire 1993; Budson and Price 2005). This chapter will concentrate on those aspects of memory that have particular neurobehavioral significance, recognizing that some important memory phenomena have no established neuroanatomic correlates. Nevertheless, much has been learned about the neuroanatomy of clinically apparent memory disorders. In precise terms, memory refers to the retention of information, whereas learning indicates the acquisition of that information. For the sake of conciseness, memory disorders will be regarded here as including learning as well as memory deficits.

Memory loss is a common complaint of older people, and sometimes younger ones as well. One obvious reason for this is the ominous threat of

Alzheimer's disease (AD), a common dementia that gradually erodes all cognitive and emotional function and presently cannot be cured. Despite the fact the AD presents with memory dysfunction in nearly every case (Chapter 12), memory complaints may represent a large number of other neurologic, psychiatric, and medical disorders and are also well recognized as a part of normal cognitive aging. Moreover, memory loss may not in fact be an accurate term for what the individual truly manifests, as failing memory is a widely used lay concept that may actually mean inattention, language disturbance, visuospatial impairment, executive dysfunction, depression, anxiety, and other cognitive or emotional problems. Detailed mental status examination can usually clarify the exact problem, whether it be in memory or some other realm.

Memory research in humans has blossomed over many decades, and now a rich, albeit somewhat confusing, classification has evolved. Terms that dominate the literature and need explanation are *declarative memory* (with its subtypes *episodic memory* and *semantic memory*), *procedural memory*, and *working memory* (Budson and Price 2005). These concepts will be discussed in the context of the following account, and all have important neuroanatomical correlates. For clinical purposes, three temporally distinct types of memory will also serve to structure the discussion. These varieties, known in the past as primary, secondary, and tertiary memory, are based on approximate but still useful distinctions in time: *immediate memory*, which functions over a period of seconds; *recent memory*, in which material is held for minutes to days; and *remote memory*, referring to memory storage over months or years. Each of these can be tested conveniently during the mental status examination, and each has meaningful neuroanatomic and clinical implications (Strub and Black 1993; Cummings and Mega 2003). This scheme is not meant to imply the irrelevance of other taxonomic categories of memory, but that the most helpful clinical information can be gathered by operating within this framework.

INATTENTION

The most obvious manifestation of memory dysfunction in a broad sense is failure of immediate memory, which signifies the inability to register information because of *inattention*. Patients with acute confusional state clearly exemplify this kind of deficit (Chapter 3). Most authorities prefer not to designate such a condition as amnesia, and indeed the primary problem is in the fundamental process of attention and not in the act of new learning. The neuroanatomy of attention as described in Chapter 3 is pertinent to this section.

A related and influential concept is *working memory*, the ability to hold information in short-term storage while permitting other cognitive operations to take place; a good example is the carry-over operation in arithmetic done without pencil and paper (Baddeley 1992, 2000). Working memory, discussed in Chapter 2 under the topic of attention, emphasizes the capacity of immediate memory to retain information so that further manipulation of this material by other cognitive processes can take place. It is thought that working memory operates via a central executive that oversees two major subsystems, phonological and visuospatial (Baddeley 1992, 2000). As might be predicted, functional neuroimaging studies have disclosed that these two subsystems correspond to left and right hemispheres, respectively (Frackowiak 1994). Closely associated with attention, it is not surprising that working memory is thought to depend on the prefrontal cortex (Goldman-Rakic 1992), specifically Brodmann areas 46 and 9 (Petrides et al. 1993). The prefrontal cortex has also been shown to be activated during the performance of neuropsychological tests that place strong demands on attention (Rezai et al. 1993).

AMNESIA

The ability to learn new information is of central importance in human life, and disorders of recent memory provide striking examples of the disability produced by impairment in this realm. *Amnesia* is the word used most often to describe this syndrome, one of the most important in behavioral neurology and general medicine.

Amnesia is not a unitary phenomenon, and it can be considered clinically in a number of ways. One familiar contrast is between failure of new learning—*anterograde amnesia*—and failure to consolidate recently formed memories—*retrograde amnesia*. Both types of amnesia are vividly demonstrated in patients with traumatic brain injury, for example, in whom there is a memory gap before the injury and a longer period thereafter during which new learning is impaired (Levin, Benton, and Grossman 1982). These deficits can accompany any amnesia; however, the anterograde component is typically more severe and incapacitating. The inability to acquire new learning as a result of anterograde amnesia is often accompanied by the interesting feature of *confabulation*, the fabrication of information in response to questioning, which has been associated with perseveration, impaired self-monitoring, and failure to inhibit incorrect responses (Shapiro et al. 1981).

A consideration of recent memory requires an explanation of declarative memory, the most relevant type of memory in this context. One of the major advances of recent decades is the characterization of declarative and procedural memory (Squire 1987; Zola-Morgan and Squire 1993; Budson and Price 2005). Declarative memory refers to the retention of facts and events encountered in experience, or what might be called the memory of "what" has occurred, and procedural memory designates the learning of skills and habits, or what

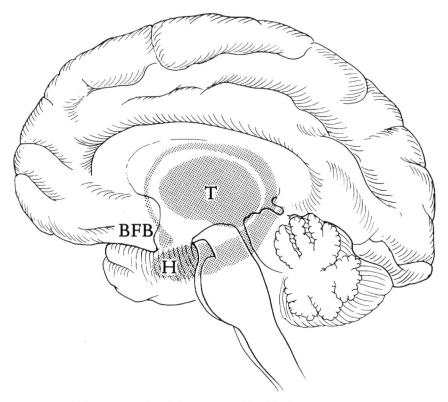

Figure **4.1.** *The hippocampus (H), thalamus (T), and basal forebrain (BFB) in relation to the right cerebral hemisphere.*

can be viewed as the memory of "how" something is done. The terms explicit and implicit memory have been used by others to describe a similar distinction, emphasizing the contrast between intentional recollection of previous experience in the former and the more automatic recall of learned skills in the latter (Schacter, Chiu, and Ochsner 1993). In clinical practice, the assessment of amnesia is ordinarily restricted to the detection of deficits in declarative memory, which is more completely understood in neurobehavioral terms, but procedural memory offers an intriguing opportunity to examine the neural basis of other types of learning that also contribute to adaptive human life.

With respect to the neuroanatomy of amnesia, the medial temporal lobe, selected nuclei of the diencephalon, and the basal forebrain are crucial (Zola-Morgan and Squire 1993). We shall consider each of these interconnected regions in turn. Figure 4.1 illustrates the relationship of these areas to the brain as a whole.

Within the temporal lobe, the hippocampus, which for our purposes will serve as shorthand for the hippocampal formation (hippocampus proper, dentate gyrus, and subiculum), has emerged as the key structure. The current understanding of amnesia can be traced to a seminal case study that has been in progress for more than fifty years and revealed more about human memory than any other project. This individual is known as H.M., a man who at age twenty-seven underwent bilateral anterior temporal lobectomy to control intractable post-traumatic epilepsy. To remove the epileptogenic cortex of the anterior temporal lobes, this operation involved resection of both the left and right hippocampi as well as the amygdalae (Scoville and Milner 1957). Despite good control of seizures achieved by the surgery, H.M. was rendered permanently amnesic by the procedure and was thereafter unable to learn any new verbal or nonverbal declarative information (Corkin 1984). Intriguingly, his ability to acquire procedural knowledge—his skill learning—was much better preserved (Milner, Corkin, and Teuber 1968; Corkin 1984), an observation that influenced the acceptance of the declarative-procedural distinction. This remarkable patient has stimulated generations of research on the role of the medial temporal lobe in recent memory, leading steadily to the acceptance of the hippocampus playing an essential role in new learning (Scoville and Milner 1957; Squire and Zola-Morgan 1991; Budson and Price 2005). With the recent death of H.M. and the results of the planned brain autopsy, more details of this extraordinary case are anticipated.

Two challenges to this "hippocampal hypothesis" have been mounted and refuted. A proposal was made by Horel (1978), who suggested that lesions of the temporal stem—the white matter tracts containing temporal lobe afferent and efferent fibers—were essential for amnesia. In the same year, Mishkin (1978) argued that combined involvement of the hippocampus and amygdala was necessary for amnesia. It seems likely, however, that hippocampal lesions are sufficient, as cases of amnesia due to anoxic injury have been reported to damage the hippocampus but spare both white matter (Cummings et al. 1984) and the amygdalae (Zola-Morgan, Squire, and Amaral 1986).

Further investigation has disclosed evidence supporting the lateralization of memory function, such that verbal declarative memory is more dependent on the left hippocampus and its connections, whereas nonverbal declarative memory is more dependent on the right side (Milner 1971; Helmstaedter et al. 1991). Such a dichotomy is in keeping with hemispheric lateralization in other areas of behavioral neurology (Chapters 5 and 8). It should be noted, however, that unilateral excision of medial temporal structures, as is often performed for intractable temporal lobe epilepsy, is rarely associated with significant amnesia (Glaser 1980). Deficits related to the side of the resection can be detected—that is, verbal impairment with left lesions and nonverbal with right—but the intact side can compensate to a considerable extent, and clinical impairment remains mild.

Two other fertile areas of clinical research on recent memory have contributed to an emerging understanding of the neuroanatomy of memory. The first is the problem of Korsakoff's amnesia—typically seen in alcoholism—in which recent memory deficits result from damage to thalamic and limbic structures anatomically connected with the medial temporal lobe. Much attention has also focused on the dementia of AD, in which memory and other cognitive deficits have been correlated with loss of neurons in the basal forebrain.

Korsakoff's psychosis, an unfortunate misnomer that has become entrenched in the medical literature, is in fact an amnesia due to deficiency of thiamine (vitamin B₁). This syndrome is usually encountered in chronic alcoholics who lack thiamine in their diet. Evolving out of the syndrome of *Wernicke's encephalopathy* (consisting of a lethargic acute confusional state, ophthalmoplegia, and gait ataxia), Korsakoff's psychosis completes the full *Wernicke-Korsakoff syndrome* (Victor, Adams, and Collins 1989). In addition to severe anterograde and variable retrograde amnesia, these patients may display confabulation, the recitation of fictitious experiences in response to an examiner's questions. Whereas the implausible, even bizarre content of many confabulated responses is often memorable and may even suggest a psychotic illness, confabulation is not a constant feature in these patients (Victor, Adams, and Collins 1989), and a more accurate name for Korsakoff's psychosis is *Korsakoff's amnesia*. Neuropathologic changes are regularly found in the thalamus and the mammillary bodies, and detailed autopsy studies have specifically emphasized the importance of dorsal medial thalamic nucleus lesions in the pathogenesis of amnesia (Victor, Adams, and Collins 1989). Supporting the validity of this diencephalic amnesia are the notable cases of N.A., a man who sustained a penetrating injury to the left dorsal medial nucleus from a miniature fencing foil and had selective verbal amnesia (Squire and Moore 1979), and A.B., a man with bilateral dorsomedial thalamic infarcts who developed dense amnesia for both verbal and figural material (Markowitsch, von Cramon, and Schuri 1993). Diencephalic amnesia related to stroke has been noted to follow infarction in the anterior but not the posterior thalamus and results from occlusion of branches of the posterior cerebral or posterior communicating arteries; careful studies have suggested that interruption of fibers connecting the dorsal medial and the anterior thalamic nuclei with the medial temporal lobe may be crucial in diencephalic amnesia (Graff-Radford et al. 1990).

The memory deficit of AD is of course the most prominent feature of in this common affliction, which will be discussed further in Chapter 12. Patients who have developed AD, with very few exceptions, present with memory loss and have clearly demonstrable amnesia on examination. Here the basal forebrain—a collection of interdigitated cell groups inferior to the basal ganglia and consisting of the medial septal nucleus, the vertical and horizontal limb nuclei of the diagonal band of Broca, and the nucleus basalis of Meynert—plays a key role (Mesulam 1990). With regard to amnesia, it seems clear that loss of cholinergic cells in the basal forebrain is both severe (Whitehouse et al. 1982) and linked with memory dysfunction in the disease (Perry et al. 1978). Basal forebrain cholinergic projections to hippocampus and neocortex are well-known (Mesulam and Geula 1988), and it is plausible that cell loss in the basal forebrain would impair memory and other cognitive functions. Evidence supporting the role of the basal forebrain in recent memory also comes from descriptions of amnesia in patients with destructive lesions of this region (Damasio et al. 1985; Morris et al. 1992).

In summary, there exists a network of structures including components of the medial temporal lobe, diencephalon, and basal forebrain that mediates the phenomenon of recent memory. Of note, positron emission tomography (PET) studies of amnesic patients with lesions of diverse etiologies showed reduced metabolism in a very similar collection of areas—the hippocampus, thalamus, cingulate gyrus, and frontal basal region—implying the existence of functional connectivity related to memory that supports the clinical findings reviewed above (Fazio et al. 1992). Figure 4.2 offers a detailed illustration of candidate regions involved in declarative memory functions. The areas depicted will be seen to overlap extensively with a group of structures known as the Papez circuit (Papez 1937), a network of interconnected cortical and subcortical structures that, in addition to participating in memory functions, has been traditionally associated with emotional behavior. The Papez circuit will be taken up in more detail in Chapter 9.

Another point concerns the interaction between the hippocampus and amygdala. Even though the hippocampus is the essential medial temporal structure for new learning, the amygdala seems to be important in the selection of material that is to be learned. Anatomic evidence indicates that these two structures are strategically positioned to intercept sensory input—visual, auditory, and tactile primarily—so that incoming stimuli can be subjected to evaluation of their biological relevance (Mesulam 1990). In brief, the amygdala functions to attach an emotional valence to information reaching the hippocampus and helps determine whether that information is worthy of storage (Chapter 9). That is, the process whereby the hippocampus either consolidates or discards information reaching the limbic system depends on the assignment of emotional significance by the amygdala.

Table 4.1 provides a listing of the diseases and injuries that affect the areas in Figure 4.1 and cause amnesia. These different etiologies should be assiduously sought and, once detected, treated promptly to maximize any recovery that might occur. In this process, it is well to bear in mind that normal aging

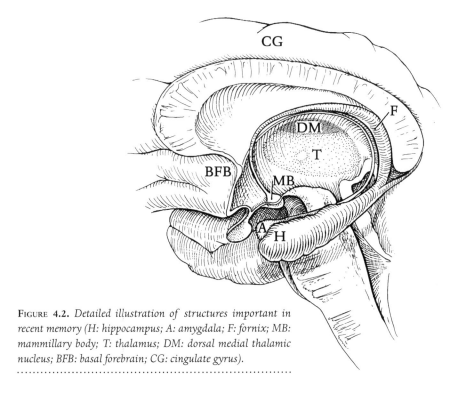

FIGURE **4.2.** *Detailed illustration of structures important in recent memory (H: hippocampus; A: amygdala; F: fornix; MB: mammillary body; T: thalamus; DM: dorsal medial thalamic nucleus; BFB: basal forebrain; CG: cingulate gyrus).*

and many psychiatric disorders can mimic neurologically based amnesia. Aging (Chapter 12) causes a mild retrieval deficit, often most marked for names, that represents an exaggeration of the normal forgetfulness shared by all humans (Devinsky 1992). Psychogenic amnesia, in contrast, usually occurs in young adults and typically features the abrupt onset of inability to recall personal information associated with an emotionally traumatic event; a good general rule is that loss of personal identity strongly implies a psychogenic, not neurologic, origin (Devinsky 1992). Thorough mental status evaluation, and neuropsychological evaluation in many cases, will usually make the necessary distinctions.

At a basic level, it is increasingly clear that the storage of memory fundamentally involves the question of how synapses change (Kandel 2006). Learning and memory likely involve synaptic plasticity, by which information of many varieties is stored in the brain. An important concept that helps explain synaptic plasticity related to memory is long-term potentiation (LTP). LTP was discovered in the rabbit hippocampus by Bliss and Lømo (1973) and has become widely accepted as the leading neurophysiologic model of learning and memory. In essence, LTP describes a lasting enhancement of a postsynaptic response

TABLE **4.1.** Etiologies of amnesia

Korsakoff's amnesia

Traumatic brain injury

Herpes simplex encephalitis

Anoxic encephalopathy

Diencephalic tumors

Posterior cerebral artery territory infarction

Transient global amnesia

Early Alzheimer's disease

Mild cognitive impairment

to incoming stimulation, implying a strengthening of synapses in the relevant neural system. The neurotransmitter glutamate is central to this process; the most abundant excitatory neurotransmitter in the brain, glutamate is an N-methyl-D-aspartate (NMDA) agonist, and NMDA receptor activation with calcium influx plays a key role in LTP (Kandel 2006). Animal studies have suggested that LTP produces altered synaptic morphology and even increased numbers of synapses, implying that synaptic density increases as a result of LTP (Kandel 2006). LTP has also been demonstrated in the human hippocampus, although access to human brain tissue is limited and samples have all come from patients who have undergone surgery for temporal lobe epilepsy (Cooke and Bliss 2006). Since LTP has been demonstrated not only in the hippocampus but also in the neocortex, it may underlie both the formation of new memories and their retention for extended periods beyond. LTP may thus represent the neurophysiologic basis of the physical changes in brain connectivity that establish each individual's unique collection of remembered experience (Kandel 2006).

REMOTE MEMORY LOSS

This division of memory is included since it represents the longest storage of experience of which the human mind is capable. An alternate word for remote memory is knowledge, and the term fund of information is also roughly equivalent. Information that is important enough to be selected for storage by the medial temporal system is encoded in the brain for future access. The precise representation of these memories in the brain has been uncertain, but the neocortex surely acts in some way as the repository for this material. Although it is unlikely that a one-to-one correspondence between a stored memory and a single cortical region exists (Squire 1987), the classic studies of neurosurgeon Wilder Penfield—who stimulated the exposed brain of awake patients undergoing epilepsy surgery—clearly demonstrated the major role of the neocortex in remote memory localization (Penfield and Jasper 1954). Animal studies have suggested that millions of widely distributed neurons are involved in the storage of even a single memory (John et al. 1986).

Knowledge of this sort implied by remote memory is usually well preserved in neurobehavioral disorders. One convincing demonstration of this principle can be seen with the syndrome of *transient global amnesia* (TGA), a reversible memory disorder affecting the hippocampus and thought to be of ischemic origin (Gonzalez-Martinez et al. 2010). In one remarkable case, a professor of music was able to play a successful organ and harpsichord concert while experiencing an episode of TGA, implying that his remote musical memories were accessible to him even as dysfunction of his recent declarative memory system rendered new learning impossible (Byer and Crowley 1980). A far more common example comes from AD patients, in whom memories from years past are relatively clear even as amnesia ineluctably precludes the storage of even the simplest new information. Well into the course of AD, a patient's birthplace, high school, first job, wedding date, and other important events can still be recalled. This phenomenon is known as Ribot's Rule, which states that remote memories are most resistant to neurologic disease (Budson and Price 2005). From these and other observations, it is apparent that a diffuse and redundant cortical representation is established as a result of frequent revisitation of important information. Nevertheless, deficits in remote memory do ultimately occur in AD with the advance of widespread cortical degeneration, at which time other signs of terminal disease—mutism, incontinence, and the vegetative state—are typically evident.

REFERENCES

Baddeley, A. The episodic buffer: a new component of working memory? *Trends Cogn Sci* 2000; 4: 417–423.

———. Working memory. *Science* 1992; 255: 556–559.

Bliss, T. V., and Lømo, T. Long-lasting potentiation of synaptic transmission in the dentate area of the anaesthetized rabbit following stimulation of the perforant path. *J Physiol* 1973; 232: 331–356.

Budson, A., and Price, B. H. Memory dysfunction. *N Engl J Med* 2005; 352: 692–699.

Byer, J. A., and Crowley, W. J. Musical performance during transient global amnesia. *Neurology* 1980; 30: 80–82.

Cooke, S. F., and Bliss, T.V.P. Plasticity in the human central nervous system. *Brain* 2006; 129: 1659–1673.

Corkin, S. Lasting consequences of bilateral medial temporal lobectomy: clinical course and experimental findings in H.M. *Semin Neurol* 1984; 4: 249–259.

Cummings, J. L., and Mega, M. S. *Neuropsychiatry and Behavioral Neuroscience.* New York: Oxford University Press; 2003.

Cummings, J. L., Tomiyasu, U., Reed, S., and Benson, D. F. Amnesia with hippocampal lesions after cardiopulmonary arrest. *Neurology* 1984; 34: 679–681.

Damasio, A. R., Graff-Radford, N. R., Eslinger, P. J., et al. Amnesia following basal forebrain lesions. *Arch Neurol* 1985; 42: 263–271.

Devinsky, O. *Behavioral Neurology: 100 Maxims.* St. Louis: Mosby Year Book; 1992.

Fazio, F., Perani, D., Gilardi, M. C., et al. Metabolic impairment in human amnesia: a PET study of memory networks. *J Cereb Blood Flow Metab* 1992; 12: 353–358.

Frackowiak, R.S.J. Functional mapping of verbal memory and language. *Trends Neurosci* 1994; 17: 109–115.

Glaser, G. H. Treatment of intractable temporal lobe–limbic epilepsy (complex partial seizures) by temporal lobectomy. *Ann Neurol* 1980; 8: 455–459.

Goldman-Rakic, P. S. Working memory and the mind. *Sci Am* 1992; 267: 111–117.

Gonzalez-Martinez, V., Comte, F., de Verbisier, D., and Carlander, B. Transient global amnesia: concordant hippocampal abnormalities on positron emission tomography and magnetic resonance imaging. *Arch Neurol* 2010; 67: 510–511.

Graff-Radford, N. R., Tranel, D., Van Hoesen, G. W., and Brandt, J. P. Diencephalic amnesia. *Brain* 1990; 113: 1–25.

Helmstaedter, C., Pohl, C., Hufnagel, A., and Elger, C. E. Visual learning deficits in non-resected patients with right temporal lobe epilepsy. *Cortex* 1991; 27: 547–555.

Horel, J. A. The neuroanatomy of memory: a critique of the hippocampal memory hypothesis. *Brain* 1978; 101: 403–445.

John, E. R., Tang, Y., Brill, A. B., et al. Double-labeled metabolic maps of memory. *Science* 1986; 233: 1167–1175.

Kandel, E. R. *In Search of Memory*. New York: W. W. Norton; 2006.

Levin, H. S., Benton, A. L., and Grossman, R. G. *Neurobehavioral Consequences of Closed Head Injury*. New York: Oxford University Press; 1982.

Markowitsch, H. J., von Cramon, D. Y., and Schuri, U. Mnestic performance profile of a bilateral diencephalic infarct patient with preserved intelligence and severe amnestic disturbances. *J Clin Exp Neuropsychol* 1993; 15: 627–652.

Mesulam, M.-M. Large-scale neurocognitive networks and distributed processing for attention, language, and memory. *Ann Neurol* 1990; 28: 597–613.

Mesulam, M.-M., and Geula, C. Nucleus basalis (Ch4) and cortical cholinergic innervation of the human brain: observations based on the distribution of acetylcholinesterase and choline acetyltransferase. *J Comp Neurol* 1988; 275: 216–240.

Milner, B. Interhemispheric differences in the localization of psychological processes in man. *Br Med Bull* 1971; 27: 272–277.

Milner, B., Corkin, S., and Teuber, H. L. Further analysis of the hippocampal amnesic syndrome: 14-year follow-up study of H.M. *Neuropsychologia* 1968; 6: 215–234.

Mishkin, M. Memory in monkeys severely impaired by combined but not by separate removal of amygdala and hippocampus. *Nature* 1978; 273: 297–298.

Morris, M. K., Bowers, D., Chatterjee, A., and Heilman, K. M. Amnesia following a discrete basal forebrain lesion. *Brain* 1992; 115: 1827–1847.

Papez, J. W. A proposed mechanism of emotion. *Arch Neurol Psychiatry* 1937; 38: 725–743.

Penfield, W., and Jasper, H. *Epilepsy and the Functional Anatomy of the Human Brain*. Boston: Little, Brown; 1954.

Perry, E. K., Tomlinson, B. E., Blessed, G., et al. Correlation of cholinergic abnormalities with senile plaques and mental test scores in senile dementia. *Br Med J* 1978; 2: 1457–1459.

Petrides, M., Alivisatos, B., Meyer, E., and Evans, A. C. Functional activation of the human frontal cortex during the performance of verbal working memory tasks. *Proc Natl Acad Sci* 1993; 90: 878–882.

Rezai, K., Andreasen, N. C., Alliger, R., et al. The neuropsychology of the prefrontal cortex. *Arch Neurol* 1993; 50: 636–642.

Schacter, D. L., Chiu, C.-Y.P., and Ochsner, K. N. Implicit memory: a selective review. *Ann Rev Neurosci* 1993; 16: 159–182.

Scoville, W. B., and Milner, B. Loss of recent memory after bilateral hippocampal lesions. *J Neurol Neurosurg Psychiatry* 1957; 20: 11–21.

Shapiro, B. E., Alexander, M. P., Gardner, H., and Mercer, B. Mechanisms of confabulation. *Neurology* 1981; 31: 1070–1076.

Squire, L. R. *Memory and Brain*. New York: Oxford University Press; 1987.

Squire, L. R., and Moore, R. Y. Dorsal thalamic lesion in a noted case of human memory dysfunction. *Ann Neurol* 1979; 6: 503–506.

Squire, L. R., and Zola-Morgan, S. The medial temporal lobe memory system. *Science* 1991; 253: 1380–1386.

Strub, R. L., and Black, F. W. *The Mental Status Examination in Neurology*. 3rd ed. Philadelphia: F. A. Davis; 1993.

Victor, M., Adams, R. D., and Collins, G. H. *The Wernicke-Korsakoff Syndrome*. 2nd ed. Philadelphia: F. A. Davis; 1989.

Whitehouse, P. J., Price, D. L., Struble, R. G., et al. Alzheimer disease and senile dementia: loss of neurons in the basal forebrain. *Science* 1982; 215: 1237–1239.

Zola-Morgan, S., and Squire, L. R. Neuroanatomy of memory. *Ann Rev Neurosci* 1993; 16: 547–563.

Zola-Morgan, S., Squire, L. R., and Amaral, D. G. Human amnesia and the medial temporal region: enduring memory impairment following a bilateral lesion limited to field CA1 of the hippocampus. *J Neurosci* 1986; 6: 2950–2967.

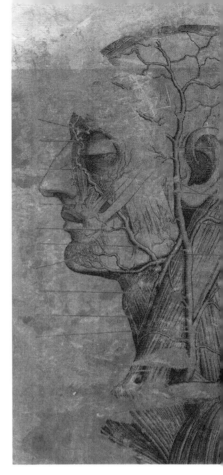

CHAPTER FIVE

LANGUAGE DISORDERS

Of all the higher functions, the capacity to communicate with language is perhaps the most obvious skill possessed by humankind. Language can be considered the verbal expression of symbolic thinking, and as such it not only endows human communication with an expansive richness of cognitive and emotional associations but also reflects the unique capacity of the human mind to interpret the world symbolically (Deacon 1997). Animals such as chimpanzees have been observed to exhibit some aspects of language, and Charles Darwin held that humans were essentially apes with big brains (Darwin 1872). But only humans possess the ability to use symbolic language to express thought, a crucial advantage likely related to enhanced synaptic density in left hemisphere regions subserving language (Premack 2007). In comparison with that of other primates, human language is extensively developed and helps confer an extraordinary mastery of the environment to which no other species can lay claim. Language is also the most thoroughly understood of all the neurobehavioral domains, primarily because of the elegant study of aphasia that ranks among the most significant achievements of behavioral neurology. The work of some of the most astute clinicians of the nervous system over the past 150 years yielded many insights into the cerebral localization of language, and these have now been established as generally valid in clinical practice (Geschwind 1970; Damasio 1992). Today, as continued investigation exploits modern neuroimaging, the classic schema of language localization has been largely confirmed, and further advances are occurring in understanding how specific cognitive processes underlying language are represented in the distributed neural networks described by behavioral neurologists (Hillis 2007).

Some definitions will serve as a point of departure. *Language*, for our purposes, will be considered the symbolic system of verbal and written communication among human beings. All cultures employ language, the primary means by which humans exchange thoughts and a source of exceptional adaptive value. It is, in fact, difficult to imagine life without language—the painful struggles of a severely aphasic patient attempting to communicate with family and friends testify to the central role of language in human interaction. A disorder of language due to brain damage is usually called *aphasia*, but the alternative dysphasia is also employed; the close similarity of dysphasia to the unrelated term dysphagia, however, is a strong argument for the use of aphasia as the preferred descriptor for an acquired language disorder. *Speech* is a more elementary capacity and refers to the mechanical act of uttering words using the neuromuscular apparatus responsible for phonation and articulation. *Dysarthria*, the term describing an impairment of speech, may be due to either brain or neuromuscular dysfunction, but it does not in itself imply a language disorder. Another term that finds utility in many settings is *voice*, the quality of speech as it is produced through the larynx. Voice disorders commonly occur as a result of laryngeal disease, such as laryngitis, and can be mild, as in *dysphonia*, or severe, as in *aphonia*. Finally, *prosody*, to be considered in Chapter 8, is a paralinguistic aspect of communication whereby language is endowed with affect.

More difficult to conceptualize is the notion of thought. Most of us know what thought is in a general sense, but since everyday thought is routinely expressed through language, can the two be separated? A recurrent philosophical issue has thus appeared, and it has been argued, for example, that only through language is thought expressed (Arendt 1978). The question is not easily answered, but clinical experience suggests that the two are indeed distinct; visual tasks demonstrating sophisticated problem solving can be completely spared in markedly aphasic patients, and, conversely, thinking may be grossly disturbed in schizophrenic patients (with a "thought disorder") who have intact language (Damasio 1992). There are also neuroanatomic reasons to suspect that thought and language are distinct; the areas responsible for language are relatively limited, whereas the remainder of the cerebrum, where nonlinguistic capacities are organized, is extensive. The neurobehavioral position is that thought, a complex and multifaceted activity, is often revealed through the symbolic system of language but cannot be regarded as synonymous with language since many other avenues for the operations of thought are available.

With these concepts in mind, we can begin to construct a neuroanatomy of language. As is true of all higher functions, no fully adequate animal model for studying language under controlled experimental conditions exists, and therefore we must turn to the distressing but revealing brain lesions that cause aphasia and related disorders. Most often, the event that disturbs language is a *stroke*; most of the history of aphasia is essentially a chronicle of the linguistic deficits consequent to focal cerebrovascular disease. Aphasia can also accompany other focal pathologies such as brain tumors, abscesses, encephalitides, traumatic contusions, and epileptogenic foci, but the detailed case study of stroke-related aphasia has led to the most notable advances in the understanding of the cerebral basis of language (Damasio 1992). An interesting development of recent years, however, has been the realization that many neurodegenerative diseases also cause aphasia, and it is clear that these cases pose different challenges because the aphasia is steadily worsening instead of improving or remaining static, as is the case with stroke-related aphasia (Hillis 2007). The neurodegenerative diseases causing aphasia are mentioned below and discussed in more detail in Chapter 12.

Although descriptions of aphasia have been made in some manner since ancient times, the modern era of aphasiology began in the late nineteenth century with the appearance of two seminal publications (Broca 1861; Wernicke 1874). In 1861, the French physician and anthropologist Paul Broca presented the case of a fifty-one-year-old man named Leborgne who had become speechless with the exception of the utterance "tan-tan." Nicknamed "Tan" because of this unfortunate condition, he had extensive destruction of the left inferior frontal lobe and adjacent areas at autopsy (Broca 1861). Broca later reported other cases that further supported the localization of speech production in the left frontal lobe (Broca 1865). Thirteen years after Broca's first case presentation, the German neurologist Carl Wernicke published his doctoral thesis, which concluded that impaired language comprehension was associated with destruction of the left superior temporal lobe (Wernicke 1874). Subsequently, other case descriptions came to light confirming these observations, and as more examples of aphasia were analyzed, a schema for left hemisphere language representation began to take shape. The contributions of Broca and Wernicke, establishing the primacy of cerebral areas essential for language in the left hemisphere, stand as superb examples of the lesion method, demonstrating how higher functions can be related to brain regions. Other investigations since that time have refined the study of aphasia, and there have been debates regarding details of the aphasia syndromes, but the localizing value of these two studies has clearly stood the test of time. The method of clinical-anatomic correlation, now made more convenient and precise by modern neuroimaging (Naeser and Hayward 1978; D'Esposito 2000; Hillis 2007), remains a powerful technique in the understanding of the brain in health as well as disease.

The literature on aphasia is extensive and often difficult. Many different classification schemes of various aphasias have been proposed, often leading to much confusion. It is not our purpose to discuss all the subtleties of aphasia as a syndrome, as others have done so in detail (Benson 1979; Albert et al. 1981; Damasio 1992; Hillis 2007), but rather to draw upon the clinical literature as it illuminates general conclusions about the neuroanatomy of language. First, however, it will prove useful to deal with the sometimes vexing issues concerning the lateralization of language.

CEREBRAL DOMINANCE AND HANDEDNESS

The crucial observation made by Broca in 1861 was that language disturbance followed damage to the left but not the right cerebral hemisphere. His case and many others since then have made it abundantly clear that, in the great majority of instances, significant aphasia follows lesions of the left hemisphere and that comparably sized and placed lesions in the right hemisphere do not as a rule cause the syndrome. It has traditionally been taught that the localization of language in the left hemisphere is true for nearly all right-handers (99 percent) and even the majority (67 percent) of left-handers (Damasio and Damasio 1992). The lateralization of language remains the most striking example of a higher function that is confined to one side of a generally symmetrical organ. The reasons for this are not immediately apparent, as other paired organs such as the lungs and kidneys manifest no such specialization. Nevertheless, knowledge that language is generally associated with the left hemisphere has important localizing value for the clinician and the neuroscientist.

The findings of Broca and his successors steadily led to a notion of *cerebral dominance* for language. The left hemisphere has been dubbed the "dominant" one, reflecting the view that language is held to be so indispensable to normal human life. Whereas the concept of dominance is frequently invoked, a certain degree of caution is warranted since, while the importance of language is indisputable, other capacities are equally vital to human existence. As will be apparent at several points throughout this book, the right hemisphere participates significantly in many aspects of attention, emotion, visuospatial skills, music, and other abilities without which our lives would be markedly impoverished. Moreover, other functions of undeniable significance, such as memory and executive function, are best regarded as bilaterally distributed. Reference to the left hemisphere as the one dominant *for language* is appropriate, but it must be realized that other important functions are associated with the right hemisphere or with both sides of the cerebrum.

As discussed above, some individuals are not left hemisphere dominant for language. The most obvious of these are people who are not right-handed or dextral. As a general rule, about 10 percent of the population is left-handed or sinistral (Tommasi 2009), and of these, perhaps a third have right hemisphere dominance for language (Damasio 1992). The situation, however, may be more complicated. Other estimates propose that 70 percent of the population are strongly right-handed, 10 percent are strongly left-handed, and 20 percent are ambidextrous; these figures have been interpreted as indicating that dextrals are left dominant for language, while sinistrals and ambidextrals have *anomalous dominance*, meaning that language is bilaterally represented (Geschwind and Behan 1984). In those with anomalous dominance, aphasia may occur following damage to either hemisphere but, when present, tends to be milder and to have a better prognosis than a similar syndrome in a dextral. The main clinical point is that variations from strong dextrality suggest the possibility that language is dominant in the right hemisphere or distributed in both. *Crossed aphasia*, the combination of aphasia and a right hemisphere lesion in a right-handed person, occurs but is extremely rare (Alexander, Fischette, and Fisher 1989).

The percentages above supporting the idea of anomalous dominance are interesting in view of neuroanatomic data indicating interhemispheric structural differences in the planum temporale, a triangular cortical area on the superior surface of the temporal lobe more or less equivalent to Wernicke's area (Figure 5.1). In a postmortem study of 100 brains, Geschwind and Levitsky (1968) found that the planum temporale is larger on the left in 65 percent of individuals, on the right in about 11 percent, and equal in the remainder. Asymmetry of the superior temporal lobe has also been shown in newborns (Witelson and Pallie 1973) and in utero (Chi, Dooling, and Gilles 1977), indicating that there appears to be a tendency for cerebral lateralization of language very early in development.

The other exception to left hemisphere dominance for language is evident in children. Despite the differences in cortical architecture described above, full lateralization of language does not appear until after childhood. The precise age when this development occurs is not clear and may vary depending on many factors, but children under ten years of age seem to have less completely lateralized language functions. Aphasias can certainly occur in children with left hemisphere lesions and often resemble adult syndromes rather closely (Cranberg et al. 1987), but recovery from seemingly devastating lesions can be quite remarkable (Smith and Sugar 1975). This resilience of children with neurological damage surely depends to a large degree on the plasticity of the young brain, which is far greater than in older people.

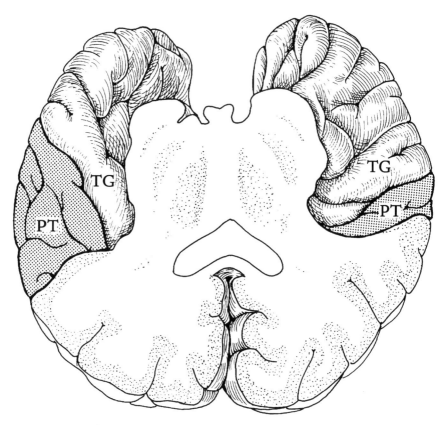

FIGURE 5.1. *Typical left-right asymmetry of the upper temporal lobes; the planum temporale is larger on the left (PT: planum temporale; TG: transverse gyrus of Heschl).*

APHASIA

Aphasia, like many neurobehavioral syndromes, can be a perplexing problem for clinicians since both the assessment of language and its disturbances are challenging topics in their own right. In many cases, the aphasic patient does not fit neatly into one of the diagnostic categories discussed below, and clinicians may come to believe that what is taught about language localization may not be as reliable as textbooks and professors claim. Whereas it is true that the wide clinical variability of aphasic patients can prove humbling, some general principles should be borne in mind. First it must be recognized that examination of linguistic functions requires considerable subjectivity; as a result, for example, impaired language fluency is best seen as a continuous rather than a dichotomous variable (Hillis 2007). Next, given the variability of brain anatomy

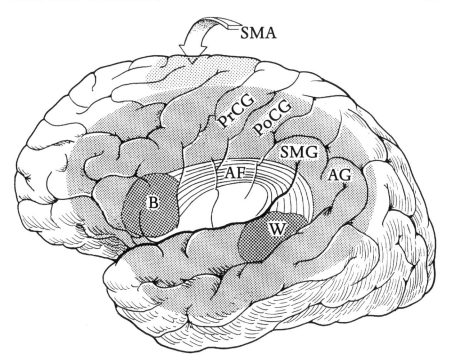

FIGURE 5.2. *Lateral view of the left hemisphere showing areas important for language (B: Broca's area; W: Wernicke's area; AF: arcuate fasciculus; SMA: supplementary motor area; AG: angular gyrus; SMG: supramarginal gyrus; PrCG: precentral gyrus; PoCG: postcentral gyrus).*

and neocortical cytoarchitectonics, the localization of language must be seen as approximate, not invariant (Hillis 2007). Last, at least in the setting of stroke, many aphasic patients have initial syndromes that are quite different from what the lasting aphasia will come to be. This is because acute cerebral lesions involve many factors that will subside in the long term, such as perilesional ischemia or edema, diaschisis, associated medical problems, medication effects, and emotional stress. Yet still, a comprehensive clinical assessment of aphasia, combined with appropriate use of structural neuroimaging and adequate follow-up, will usually reveal that the localization of language holds up in clinical practice, and the specific aphasia diagnosis will prove useful.

The collection of gray and white matter structures around the Sylvian fissure in the left hemisphere is perhaps the best-known region in behavioral neurology. This area, together with the adjacent neocortex extending from it superiorly and posteriorly, constitutes the cerebral basis of propositional language (Geschwind 1965; Damasio 1992; Hillis 2007; Figure 5.2). Reading and writing, as will be seen, also depend on this neuroanatomic region, with contributions from the visual system, and a recently identified neural network for spelling is also based in these areas (Cloutman et al. 2009). The breakdown in all of these interrelated linguistic capacities can be studied with the use of a well-known clinical instrument known as the Boston Diagnostic Aphasia Examination (BDAE; Goodglass and Kaplan 1983), which will provide much of the basis for the account that follows. Prosody, the term generally referring to the affective features of language as opposed to its propositional components, is mediated by an analogous zone on the right side of the cerebrum (Ross 1981; Ross and Monnot 2008); we will deal with this topic and the aprosody syndromes in Chapter 8.

The area of the cerebrum devoted to the production of language is Broca's area, located in the posterior portion of the left inferior frontal gyrus (Figure 5.2). This is essentially the same territory occupied by Brodmann areas 44 and 45. Here the semantic components of a phrase or sentence are assembled into a grammatically correct and well-articulated utterance. Lesions of this area are associated with often dramatic difficulties in verbal output, an impairment appropriately referred to as nonfluent speech. Severe, lasting *Broca's aphasia* is due to damage involving Broca's area and adjacent cortical and subcortical regions (Alexander, Naeser, and Palumbo 1990), whereas a milder "Broca's area aphasia" results from damage to Broca's area exclusively (Mohr et al. 1978). *Aphemia*, a nonaphasic syndrome of transient mutism or dysarthria, has also been described with small lesions of the lower precentral gyrus that spare Broca's area (Schiff et al. 1983). These syndromes are usually due to a stroke in the distribution of the left middle cerebral artery, but nonfluent aphasia from neurodegenerative disease—initially described as the syndrome of *primary progressive aphasia*—has also been observed (Mesulam 1982; Weintraub, Rubin, and Mesulam 1990). A patient with Broca's aphasia, or any nonfluent aphasia, is not unlike that of a nonaphasic person struggling with a foreign language; speech is typically sparse, effortful, and grammatically incorrect. Because of the proximity of Broca's area with the lower portions of the precentral gyrus, weakness or paresis of the contralateral (right) arm and face is usually seen with Broca's aphasia; sensory loss in the same bodily regions is sometimes present as well if the postcentral gyrus is involved. Patients with this syndrome have relatively preserved auditory comprehension, since the production of language is primarily affected.

In contrast, the region of the cerebrum most concerned with the understanding of language is situated in the posterior portion of the left superior temporal gyrus (Figure 5.2). This is known as Wernicke's area and corresponds with the posterior part of Brodmann area 22. Wernicke's area is responsible for processing incoming verbal stimuli into linguistically meaningful information. Damage to this region leads to impaired auditory comprehension despite retained fluency of language, a syndrome known as *Wernicke's aphasia* (Geschwind 1967).

As with Broca's aphasia, a left middle cerebral artery distribution stroke is the usual cause of Wernicke's aphasia, but similar fluent aphasias frequently occur as a component of Alzheimer's disease (AD; Weintraub, Rubin, and Mesulam 1990; Chapter 12). A patient with Wernicke's aphasia has difficulty understanding spoken language, including his or her own as well as that of others. In contrast, speech production is fluent, although typically rapid and paraphasic. The term *logorrhea* is not inappropriate in such cases, and the *press of speech*—riddled with literal or semantic paraphasias and uninformed by adequate auditory comprehension—may be impressive. Paraphasias may also take the form of neologisms, and when neologistic speech is severe, the term *jargon aphasia* is sometimes used. No paresis or somatosensory loss is typically present in patients so afflicted, and the syndrome of Wernicke's aphasia is the conceptual opposite of Broca's aphasia in that language input is affected while its output is better preserved.

Broca's and Wernicke's areas are also connected by a subcortical white matter tract known as the arcuate fasciculus (Figure 5.2). This structure is important in that the act of repetition is regarded by many as dependent on its integrity. As originally predicted by Wernicke (1874) and later supported by Geschwind (1965), damage to the arcuate fasciculus, typically with associated cortical pathology, results in selective impairment of repetition; there is an interruption of the transfer of accurately processed input in Wernicke's area to the intact output zone of Broca's area, and repetition of sentences, phrases, or even single words breaks down, often with paraphasic word or letter substitutions. This syndrome is called *conduction aphasia* (Damasio and Damasio 1980), and affected patients display a special deficit in repetition that may be quite surprising in the presence of their spared auditory comprehension and verbal fluency. Whereas many neurologists accept the role of the arcuate fasciculus in conduction aphasia, some question this localization in view of reported cases of conduction aphasia that appear to have lesions in other areas, such as the overlying parietal neocortex or the deep parietal white matter (Hillis 2007).

So far, we have seen the effects of limited focal lesions in the language zone, but there is an unfortunate syndrome involving the entire perisylvian region. When a large infarct destroys the areas discussed above, which is sadly not infrequent, the deficits are additive, and individuals suffer with impaired language output, input, and of course repetition. This *global aphasia*, accompanied by hemiplegia of the right side of the body, is one of the more tragic sequelae of central nervous system disease. Efforts to prevent cerebrovascular disease are in no setting more justified than in the quest to avert the onset of global aphasia. As can be imagined, the outcome is typically poor. A better prognosis in global aphasia has been seen, however, when two separate focal embolic lesions in Broca's and Wernicke's areas occur, a condition known as *global aphasia without hemiparesis* (Tranel et al. 1987).

A final important component of spoken language is naming. The ability to name an object—for example, the collection of pages the reader is holding known as a book—depends on the assignment of a verbal symbol to an object that, in this case, is visually perceived. The resultant name that is generated becomes, of course, one of the foundations of language, permitting the development of a lexicon or vocabulary, and the expansion of symbolic representation. The power of language as conveyed by prose, poetry, academic or scientific writing, and ordinary discourse stems from the capacity of words to impart a panoply of symbolic meanings, extending language beyond the denotative to the connotative. The left angular gyrus (Brodmann area 39) is anatomically placed to receive auditory, visual, and somatosensory information and has been suggested as the region where naming takes place by virtue of cross-modal associations (Geschwind 1965). Lesions in this area are regularly associated with prominent anomia, and the syndrome of *anomic aphasia* is characteristically seen with damage to the left angular gyrus. In fact, however, anomia is a universal feature of aphasia and can therefore be caused by lesions that cause any aphasic syndrome. Furthermore, anomia is often encountered in diffuse conditions such as acute confusional state and AD, in which word-finding difficulty is embedded within the many other deficits present. Despite its poor localizing value in many cases, further study of anomia may reveal subtypes that can be linked more specifically with known language areas.

We have thus far described the linguistic functions of the perisylvian zone in the left hemisphere. There remains a large expanse of cortex and subjacent white matter that abuts upon this area (Figure 5.2). Although less familiar than the major language areas discussed above, these extrasylvian zones of the brain contribute importantly to normal language, and the study of syndromes related to their destruction is instructive. Four aphasia syndromes have been associated with this region of the left hemisphere (Table 5.1). These aphasias have all been grouped under the heading of *transcortical aphasia*, an older term with an imprecise meaning; even though the transcortical terminology persists, considering these syndromes under the rubric of *extrasylvian aphasia* is more accurate.

As a first point, these aphasias all demonstrate sparing of repetition, which is not surprising because the primary circuit mediating auditory input, speech output, and the connections between them is intact. Extrasylvian aphasias are typically due to hypotensive or hypoxic injury in the borderzone regions of the hemisphere where terminal anastomoses occur between the middle cerebral and the anterior and posterior cerebral arteries. This neurovascular feature accounts for the preservation of the perisylvian region, which is irrigated solely by the middle cerebral artery. The first syndrome is anomic aphasia, which has been considered above and is often due to damage in the left angular gyrus posterior to the Sylvian fissure. *Transcortical motor aphasia* is due to a lesion in

TABLE **5.1.** Aphasia syndromes and their localization

Aphasia Type	Spontaneous Speech	Auditory Comprehension	Repetition	Naming	Localization (Left Hemisphere)
Broca's	Nonfluent	Good	Poor	Poor	Broca's area
Wernicke's	Fluent	Poor	Poor	Poor	Wernicke's area
Conduction	Fluent	Good	Poor	Poor	Arcuate fasciculus
Global	Nonfluent	Poor	Poor	Poor	Perisylvian region
Transcortical motor	Nonfluent	Good	Good	Poor	Anterior borderzone
Transcortical sensory	Fluent	Poor	Good	Poor	Posterior borderzone
Anomic	Fluent	Good	Good	Poor	Angular gyrus
Mixed trans-cortical	Nonfluent	Poor	Good	Poor	Anterior and posterior borderzone

the anterior extrasylvian region, often involving the supplementary motor area (Brodmann area 6), and characterized by particular difficulty in the initiation of speech (Alexander and Schmitt 1980; Freedman, Alexander, and Naeser 1984). *Transcortical sensory aphasia,* marked by difficulty with language comprehension, follows lesions in the posterior extrasylvian region (Brodmann area 37 and portions of 39 and 19) and their deep white matter connections (Kertesz, Sheppard, and MacKenzie 1982; Alexander, Hiltbrunner, and Fischer 1989). These two aphasias are analogous to Broca's and Wernicke's aphasias, with the notable exception of spared repetition. Finally, the intriguing entity of *mixed transcortical aphasia,* or "isolation of the speech area," occurs and features damage to the entire C-shaped extrasylvian region; then the only remaining skill is repetition (Geschwind, Quadfasel, and Segarra 1968). Individuals with this aphasia may demonstrate the curious phenomenon of *echolalia,* in which the examiner's statements are automatically repeated. Table 5.1 displays the modern classification of these eight major aphasias, including the various deficits by which they are characterized and the affected cerebral areas.

Before leaving the topic of spoken language, the category of *subcortical aphasia* deserves comment. The reported syndromes of aphasia due to subcortical damage, usually ischemic, have been controversial since the cortex is traditionally seen as responsible for language function, but clearly documented aphasias have been reported from lesions of the left thalamus, basal ganglia, and white matter (Filley and Kelly 1990). In general, these are relatively mild syndromes that resemble the transcortical aphasias in that repetition is spared. The prognosis is usually favorable. Some have argued that subcortical aphasia is due simply to transient metabolic alterations in overlying language cortex (Olsen, Bruhn, and Oberg 1986), but alternatively, there may be subcortical components of language systems that serve to "activate" the cortex in the production of its linguistic activities. It is premature to be certain of the contributions of the subcortical regions to language, but further case studies will be useful in this regard.

ALEXIA

In illiterate individuals, reading and writing are of course poorly or not at all developed. However, in patients with brain damage who have acquired these skills, much has been learned of the cerebral origin of these linguistic capacities. Two lines of inquiry have established most of our present knowledge of the anatomy of reading: the study of acquired disorders of reading—alexia— and the investigation of developmental dyslexia, a syndrome in which children may never learn to read well despite ample opportunity and general cognitive ability.

Alexia has been classically divided into two distinct varieties: alexia with agraphia and alexia without agraphia (pure alexia). Both syndromes were recognized in the late nineteenth century, and, like many other neurobehavioral syndromes of that era, both have endured through a century of clinical research. We shall consider these syndromes and a third alexia added more recently.

The French neurologist Jules Dejerine described alexia with agraphia and pure alexia in two publications appearing a year apart (Dejerine 1891, 1892). *Alexia with agraphia* is a disorder of both reading and writing that can be regarded as a syndrome of acquired illiteracy. The lesion involves the inferior parietal and posterolateral temporal regions of the left hemisphere, most critically the angular gyrus (Dejerine 1891), and associated deficits including right homonymous visual field defects and mild fluent aphasia, and components of Gerstmann's syndrome may be present (Figure 5.2). Involvement of the angular gyrus in this syndrome is of interest in view of the likelihood that this gyrus plays a special role in linguistic function. Similar to its role in the process of naming, the angular gyrus probably enables the cross-modal associations between visual and auditory systems by which one learns to read (Geschwind 1965).

Alexia without agraphia (Dejerine 1892), on the other hand, serves as an instructive contrast to alexia with agraphia because in this syndrome reading is selectively affected and the angular gyrus is intact. An excellent example of a *disconnection syndrome* (Geschwind 1965), alexia without agraphia—or pure alexia, as it is sometimes known—is typically due to lesions in the left occipital lobe and the splenium of the corpus callosum that prevent incoming visual information

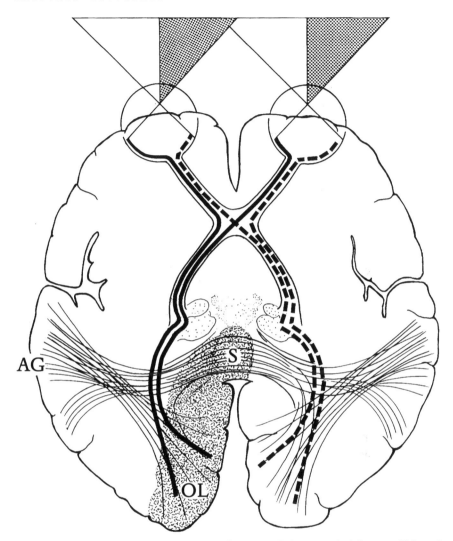

FIGURE 5.3. *Horizontal section of the cerebrum illustrating the lesions in the left occipital lobe and the splenium of the corpus callosum resulting in alexia without agraphia (OL: occipital lobe; S: splenium of the corpus callosum; AG: angular gyrus).*

from reaching the left angular gyrus for linguistic interpretation (Figure 5.3). Thus one can observe the remarkable phenomenon of a patient who can write easily but has difficulty reading what he or she has just written. Right homonymous hemianopia is characteristically present, and color anomia and associative visual agnosia can be seen as well.

Finally, another variety of reading disturbance has been called *frontal alexia*, alternatively known as anterior alexia because of its association with Broca's aphasia or the "third alexia" (Benson 1977). Patients with this syndrome have severe nonfluent aphasia and right hemiparesis, and their reading impairment is characterized by inability to read letters despite preserved reading of familiar words (literal alexia). In addition, words with high semantic content such as "house" and "car" are read more easily than small functor words such as "the" and "in," so that the reading disturbance parallels the spoken language of patients with Broca's aphasia.

The alexias are an intriguing group of syndromes that offer insights into the complicated cerebral processes of reading. For clinical purposes, the most important point is that acquired reading disorders can often be overlooked unless specifically sought. Neuroanatomically, it is convenient to envision the alexias as left hemisphere syndromes related to posterior (alexia without agraphia), central (alexia with agraphia), and anterior (frontal alexia) pathology. The precise contributions that each of these areas makes to the many steps involved in reading await further study.

Another form of reading disorder has long been recognized, and because it affects large numbers of otherwise normal children, this disorder has rightly captured much popular attention. *Developmental dyslexia*, or difficulty learning to read despite adequate intelligence, motivation, and educational opportunity, is a common disorder, estimated to occur in 5–10 percent of school-age children (Shaywitz and Shaywitz 2008). More common in boys than girls, dyslexia is an anomaly of left hemisphere development with both genetic and environmental determinants (Pennington 1991). The specific neuroanatomy of the impairment in reading and spelling is under investigation, but some evidence suggests that the planum temporale (Figure 5.1) is part of the cerebral network necessary for the normal development of reading. Dyslexic children often show loss of the normal left-right asymmetry in the planum region, so that the plana are symmetric in size (Galaburda et al. 1985). Positron emission tomography studies have also found metabolic dysfunction in the left temporoparietal region of dyslexic men (Rumsey et al. 1994). Microscopic abnormalities such as neuronal ectopias and architectonic dysplasias in the perisylvian zone of the left hemisphere have also been seen as contributing to dyslexia (Galaburda et al. 1985). More recent structural neuroimaging has disclosed left parietotemporal white matter changes in dyslexic subjects (Klingberg et al. 2000), and functional neuroimaging has identified abnormalities in left parietotemporal and occipitotemporal regions (Shaywitz and Shaywitz 2008). In keeping with the remarkable prescience of classic neurologists, these data are generally consistent with the initial formulation of Dejerine on the cerebral localization of reading (Dejerine 1891, 1892). These recent findings also support results of cognitive research suggesting that dyslexia is a disorder of written language processing, not a visual or spatial processing problem (Pennington 1991). A consensus view is emerging on what has come to be called the phonological model of dyslexia, which holds that reading requires the linkage of orthography to phonology—translating letters to sounds—and that developmental disruption of posterior left hemisphere structures limits the capacity of affected children to accomplish this task (Shaywitz and Shaywitz 2008).

AGRAPHIA

Writing, like reading, is a linguistic skill not possessed by all people with a command of spoken language. Even in those who are literate, writing is employed less often than spoken language for communication and is therefore less securely represented in the brain. *Agraphia*, the impairment of previously intact writing due to brain damage, can thus be caused by lesions virtually anywhere in the cerebrum. Indeed, the fragility of writing in disorders of the brain is evidenced by the fact that agraphia routinely appears even in diffuse syndromes such as the acute confusional state (Chedru and Geschwind 1972) and dementia (Appell, Kertesz, and Fisman 1982). This section will consider the various forms in which writing may break down; the curious phenomenon of hypergraphia will be taken up in Chapter 9.

On the basis of this syndrome appearing with a variety of lesions, many classifications of agraphia have been attempted (Roeltgen 2003), none entirely successful. It seems reasonable, however, to begin with the observation that a number of agraphia syndromes can be ascribed to motor and even non-neurologic disorders, a group aptly named the mechanical agraphias (Benson 1994). In this category would be writing impairments due to corticospinal dysfunction, basal ganglia disease, cerebellar disturbance, peripheral neuropathy, myopathy, and bone or joint disease. Neurologists are familiar with the clumsy writing resulting from paresis, the micrographia of Parkinson's disease, and the tremulous writing due to cerebellar tremor. These deficits can be demonstrated with the elemental neurological examination and do not reflect a cognitive disorder, but they often coexist with neurobehavioral syndromes and cloud the clinical picture.

The localization of writing has not been resolved with certainty. As might be imagined, most often agraphia accompanies aphasia following left hemisphere lesions, and these tend to cluster in the perisylvian region (Roeltgen 2003). Typically the features of the agraphia tend to parallel the characteristics of the aphasia; as examples, agraphia with nonfluent aphasia shows effortful and sparse written output with agrammatism, whereas agraphia with fluent aphasia discloses easily written output of normal length but with uncertain meaning and paragraphic errors. Spatial agraphia with right hemisphere lesions is also recognized, in which errors are made due to hemineglect and constructional impairments (Ardila and Rosselli 1993). Agraphia can also be seen in the left hand only when a patient has a lesion of the corpus callosum (Geschwind 1965).

In view of the many clinical settings giving rise to agraphia, cases that occur in the absence of other disorders are of considerable interest. Such examples, unfortunately, are uncommon. Rare cases of pure agraphia have been reported with lesions of the left superior frontal region (Dubois, Hecaen, and Marcie 1969), of the left superior parietal lobe (Auerbach and Alexander 1981), and in "Exner's area"—the left premotor region immediately anterior to the hand region of the precentral gyrus (Exner 1881). In practice, however, agraphia is most often commingled with other neurobehavioral deficits (Roeltgen 2003).

Agraphia is also one member of a tetrad known as *Gerstmann's syndrome*, the others being acalculia, finger agnosia, and right-left disorientation (Gerstmann 1940). This aggregate of clinical features has been claimed to localize a lesion to the left angular gyrus (Gerstmann 1940). The localizing value of the entity has been alternatively questioned (Benton 1961) and affirmed (Strub and Geschwind 1974). Whereas demonstration of the full syndrome may at times implicate the left angular gyrus, detection of partial Gerstmann's syndrome—common in clinical practice—cannot be used to identify a left angular gyrus lesion. The four components of Gerstmann's syndrome may seem a peculiar collection, but intriguing links exist. First, the problem of *acalculia* deserves comment. As with reading and writing, the ability to calculate is a skill mainly confined to persons with some educational background, and disorders of writing are seen in a wide range of developmental, vascular, and degenerative diseases (Grafman and Rickard 1997). When an educated individual develops an isolated impairment in the performance of arithmetic operations, the term *anarithmetria* is appropriate and a left parietal lesion can be suspected (Takayama et al. 1994; Grafman and Rickard 1997). Acalculia can in turn be linked to the other two components of Gerstmann's syndrome by recalling that preliterate calculation involves using all fingers of both hands. Thus all four elements of Gerstmann's syndrome can be associated with the left angular gyrus. Although the debate over the clinical utility of Gerstmann's syndrome will likely continue, the left angular gyrus is nevertheless an important one for behavioral neurology. Large lesions in this area have been noted to cause the *angular gyrus syndrome*—anomic aphasia, alexia with agraphia, and Gerstmann's syndrome—that may closely mimic the deficits of AD (Benson, Cummings, and Tsai 1982).

REFERENCES
....

Albert, M. L., Goodglass, H., Helm, N. A., et al. *Clinical Aspects of Dysphasia*. Vienna: Springer-Verlag; 1981.

Alexander, M. P., Fischette, M. R., and Fisher, R. S. Crossed aphasia can be mirror image or anomalous: case reports, review and hypothesis. *Brain* 1989; 112: 953–973.

Alexander, M. P., Hiltbrunner, B., and Fischer, R. S. Distributed anatomy of transcortical sensory aphasia. *Arch Neurol* 1989; 46: 885–892.

Alexander, M. P., Naeser, M. A., and Palumbo, C. Broca's area aphasias: aphasia after lesions including the frontal operculum. *Neurology* 1990; 40: 353–362.

Alexander, M. P., and Schmitt, M. A. The aphasia syndrome of stroke in the left anterior cerebral artery territory. *Arch Neurol* 1980; 37: 97–100.

Appell, J., Kertesz, A., and Fisman, M. A study of language functioning in Alzheimer's disease. *Brain Lang* 1982; 17: 73–91.

Ardila, A., and Rosselli, M. Spatial agraphia. *Brain Cognition* 1993; 22: 137–147.

Arendt, H. *The Life of the Mind*. New York: Harcourt Brace Jovanovich; 1978.

Auerbach, S. H., and Alexander, M. P. Pure agraphia and unilateral optic ataxia associated with a left superior parietal lobule lesion. *J Neurol Neurosurg Psychiatry* 1981; 44: 430–432.

Benson, D. F. *Aphasia, Alexia, and Agraphia*. New York: Churchill Livingstone; 1979.

———. *The Neurology of Thinking*. New York: Oxford University Press; 1994.

———. The third alexia. *Arch Neurol* 1977; 34: 327–331.

Benson, D. F., Cummings, J. L., and Tsai, S. I. Angular gyrus syndrome simulating Alzheimer's disease. *Arch Neurol* 1982; 39: 616–620.

Benton, A. L. The fiction of the "Gerstmann" syndrome. *J Neurol Neurosurg Psychiatry* 1961; 24: 176–181.

Broca, P. Remarques sur le siège de la faculté du langage articulé, suives d'une observation d'aphémie. *Bull Soc Anat* 1861; 36: 333–357.

———. Sur la faculté du langage articulé. *Bull Soc Anthrop (Paris)* 1865; 6: 337–393.

Chedru, F., and Geschwind, N. Writing disturbances in acute confusional states. *Neuropsychologia* 1972; 10: 343–354.

Chi, J. G., Dooling, E. C., and Gilles, F. H. Gyral development of the human brain. *Ann Neurol* 1977; 1: 86–93.

Cloutman, L., Gingis, L., Newhart, M., et al. A neural network critical for spelling. *Ann Neurol* 2009; 66: 249–253.

Cranberg, L. D., Filley, C. M., Alexander, M. P., and Hart, E. J. Acquired aphasia in childhood: clinical and CT investigations. *Neurology* 1987; 37: 1165–1172.

Damasio, A. R. Aphasia. *N Engl J Med* 1992; 326: 531–539.

Damasio, A. R., and Damasio, H. Brain and language. *Sci Am* 1992; 267: 89–95.

Damasio, H., and Damasio, A. R. The anatomical basis of conduction aphasia. *Brain* 1980; 103: 337–350.

Darwin, C. *The Expression of Emotion in Animals and Man*. New York: Philosophical Library; 1872.

Deacon, T. W. *The Symbolic Species: The Co-evolution of Language and the Brain*. New York: W. W. Norton and Company; 1997.

Dejerine, J. Contribution à l'étude anatomo-pathologique et clinique des différentes variétés de cécité verbale. *Mem Soc Biol* 1892; 4: 61–90.

———. Sur un cas de cécité verbale avec agraphie, suivi d'autopsie. *Mem Soc Biol* 1891; 3: 197–201.

D'Esposito, M. Functional neuroimaging of cognition. *Semin Neurol* 2000; 20: 487–498.

Dubois, J., Hecaen, H., and Marcie, P. L'Agraphie "pure." *Neuropsychologia* 1969; 7: 271–286.

Exner, S. *Untersuchungen über die Lokalisation der Funktionen in der Grosshirnrinde des Menschen*. Vienna: Wilhelm Braumuller; 1881.

Filley, C. M., and Kelly, J. P. Neurobehavioral effects of focal subcortical lesions. In: Cummings, J. L., ed. *Subcortical Dementia*. New York: Oxford University Press; 1990: 59–70.

Freedman, M., Alexander, M. P., and Naeser, M. A. Anatomic basis of transcortical motor aphasia. *Neurology* 1984; 34: 409–417.

Galaburda, A. M., Sherman, G. F., Rosen, G. D., et al. Developmental dyslexia: four consecutive patients with cortical anomalies. *Ann Neurol* 1985; 18: 222–233.

Gerstmann, J. Syndrome of finger agnosia, disorientation for right and left, agraphia and acalculia. *Arch Neurol Psychiatry* 1940; 44: 398–408.

Geschwind, N. Disconnexion syndromes in animals and man. *Brain* 1965; 88: 237–294, 585–644.

———. The organization of language and the brain. *Science* 1970; 170: 940–944.

———. Wernicke's contribution to the study of aphasia. *Cortex* 1967; 3: 449–463.

Geschwind, N., and Behan, P. O. Laterality, hormones, and immunity. In: Geschwind, N., Galaburda, A. M., eds. *Cerebral Dominance: The Biological Foundations*. Cambridge, MA: Harvard University Press; 1984: 211–224.

Geschwind, N., and Levitsky, W. Human brain: left-right asymmetry in temporal speech region. *Science* 1968; 161; 186–187.

Geschwind, N., Quadfasel, F. A., and Segarra, J. M. Isolation of the speech area. *Neuropsychologia* 1968; 6: 327–340.

Goodglass, H., and Kaplan, E. *The Assessment of Aphasia and Related Disorders*. 2nd ed. Philadelphia: Lea and Febiger; 1983.

Grafman, J., and Rickard, T. Acalculia. In: Feinberg, T. E., and Farah, M. J., eds. *Behavioral Neurology and Neuropsychology*. New York: McGraw-Hill; 1997: 219–226.

Hillis, A. E. Aphasia: progress in the last quarter of a century. *Neurology* 2007; 69: 200–213.

Kertesz, A., Sheppard, A., and MacKenzie, R. Localization in transcortical sensory aphasia. *Arch Neurol* 1982; 39: 475–478.

Klingberg, T., Hedehus, M., Temple, E., et al. Microstructure of temporo-parietal white matter as a basis for reading ability: evidence from diffusion tensor magnetic resonance imaging. *Neuron* 2000; 25: 493–500.

Mesulam, M.-M. Slowly progressive aphasia without generalized dementia. *Ann Neurol* 1982; 11: 592–598.

Mohr, J. P., Pessin, M. S., Finkelstein, S., et al. Broca aphasia: pathologic and clinical. *Neurology* 1978; 28: 311–324.

Naeser, M. A., and Hayward, R. W. Lesion localization in aphasia with cranial computed tomography and the Boston Diagnostic Aphasia Exam. *Neurology* 1978; 28: 545–551.

Olsen, T., Bruhn, T., and Oberg, G. E. Cortical hypoperfusion as a possible cause of "subcortical aphasia." *Brain* 1986; 109: 393–410.

Pennington, B. F. *Diagnosing Learning Disorders: A Neuropsychological Framework.* New York: Guilford; 1991.

Premack, D. Human and animal cognition: continuity and discontinuity. *Proc Nat Acad Sci* 2007; 104: 13861–13867.

Roeltgen, D. P. Agraphia. In: Heilman, K. M., and Valenstein, E., eds. *Clinical Neuropsychology.* 4th ed. New York: Oxford University Press; 2003: 126–145.

Ross, E. D. The aprosodias: functional-anatomic organization of the affective components of language in the right hemisphere. *Arch Neurol* 1981; 38: 561–569.

Ross, E. D., and Monnot, M. Neurology of affective prosody and its functional-anatomic organization in right hemisphere. *Brain Lang* 2008; 104: 51–74.

Rumsey, J. M., Zametkin, A. J., Andreason, P., et al. Normal activation of frontotemporal language cortex in dyslexia, as measured with oxygen 15 positron emission tomography. *Arch Neurol* 1994; 51: 27–38.

Schiff, H. B., Alexander, M. P., Naeser, M. A., and Galaburda, A. M. Aphemia: clinical-anatomic correlations. *Arch Neurol* 1983; 40: 720–727.

Shaywitz, S. E., and Shaywitz, B. A. Paying attention to reading: the neurobiology of reading and dyslexia. *Dev Psychopathol* 2008; 20: 1329–1349.

Smith, A., and Sugar, O. Development of above-normal language and intelligence 21 years after left hemispherectomy. *Neurology* 1975; 25: 813–818.

Strub, R., and Geschwind, N. Gerstmann's syndrome without aphasia. *Cortex* 1974; 10: 378–387.

Takayama, Y., Sugishita, M., Akiguchi, I., and Kimura, J. Isolated acalculia due to left parietal lesion. *Arch Neurol* 1994; 51: 286–291.

Tommasi, L. Mechanisms and functions of brain and behavioural asymmetries. *Philos Trans R Soc Lond B Biol Sci* 2009; 364: 855–859.

Tranel, D., Biller, J., Damasio, H., et al. Global aphasia without hemiparesis. *Arch Neurol* 1987; 44: 304–308.

Weintraub, S., Rubin, N. P., and Mesulam, M.-M. Primary progressive aphasia: longitudinal course, neuropsychological profile, and language features. *Arch Neurol* 1990; 47: 1329–1335.

Wernicke, C. *Der aphasiche Symptomencomplex.* Breslau: Cohn and Weigert; 1874.

Witelson, S. F., and Pallie, W. Left hemisphere specialization for language in the newborn: neuroanatomical evidence of asymmetry. *Brain* 1973; 96: 641–646.

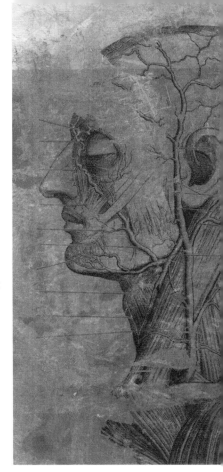

CHAPTER SIX

CHAPTER SIX

APRAXIA

Apraxia is an acquired disorder of skilled purposeful movement (Heilman and Rothi 2003; Greene 2005). As with aphasia and dysphasia, the term dyspraxia is sometimes used to denote less severe apraxia, but like aphasia, apraxia is far more commonly used for all degrees of severity. The theoretical importance of apraxia relates to the phenomenon of tool use, one of the key developments in human evolution that contributed to enhanced adaptation to the physical environment (Lewis 2006). In the clinical setting, apraxia can pose a significant impediment to patients with stroke and other brain lesions as they attempt to resume a normal life (Hanna-Pladdy, Heilman, and Foundas 2003). Diminished capacity to use tools such as pencils, forks, and toothbrushes in daily living may cause patients to experience a less favorable outcome in rehabilitation and potential lasting motor system disability. The term apraxia is often used as shorthand for ideomotor apraxia (see below), the most common and clinically important variety of the syndrome, but apraxia is best considered a generic designation for cognitive motor impairment. Thus apraxia represents a loss of motor skill qualitatively distinct from paresis much as aphasia is distinct from dysarthria. In clinical practice, an apraxic patient must be shown to have a disorder of skilled movement not caused by significant paresis, akinesia, ataxia, sensory loss, inattention, poor comprehension, or other cognitive impairment. Apraxia is commonly encountered with aphasia, but it is not simply an aspect of linguistic dysfunction; the concurrence of the two syndromes is more likely a result of the neuroanatomic proximity of the language and higher motor systems than a manifestation of a common underlying mechanism. Like aphasia, however, apraxia is typically caused by ischemic cerebrovascular disease. In essence, apraxia is a syndrome of higher motor dysfunction in the same way that agnosia (Chapter 7) represents higher sensory impairment.

Some terminological issues need to be dealt with as a first step. Many neurobehavioral syndromes with some component of motor dysfunction have been given the name apraxia, but it is questionable whether they should be considered apraxias as defined above. *Constructional apraxia* is an alternate term for visuospatial impairment, and *dressing apraxia* denotes difficulty with dressing; both are right hemisphere syndromes (Chapter 8). *Ocular apraxia*, also called oculomotor apraxia or psychic paralysis of gaze, is a component of Balint's syndrome and is described with the visual agnosias (Chapter 7). *Gait apraxia* (Denny-Brown 1958) is a disorder of gait, often called magnetic gait, seen in diseases affecting the frontal lobe white matter such as normal pressure hydrocephalus (Chapter 12). *Verbal apraxia*, or apraxia of speech, is a term sometimes used to describe impaired speech that falls intermediate between dysarthria and aphasia (Kirshner 2002). The discussion of apraxia below will confine itself to relatively uncontroversial uses of the term.

One of the most challenging syndromes in behavioral neurology, apraxia has presented theoretical problems since first recognized in the nineteenth century (Heilman and Rothi 2003). The seminal papers of Hugo Liepmann in the early 1900s (Liepmann 1900, 1906, 1908, 1920; Liepmann and Maas 1907) structured the consideration of apraxia but raised many issues that are far from resolved. One problem has been the bewildering array of definitions of apraxia, engendering much confusion about the true meaning of the term; this conceptual difficulty is illustrated by the fact that the definition advanced above actually describes what apraxia is not rather than what it is (Heilman and Rothi 2003). Furthermore, apraxic patients rarely complain of their deficits (Geschwind 1975), in contrast to the often obvious problems of those with aphasia and other syndromes. For these reasons, apraxia testing is often omitted from the mental status examination, and useful diagnostic and prognostic information is not obtained. We will consider apraxia in the context of the original classification of Liepmann, one that has enjoyed widespread use and that still serves to organize current thinking on this controversial subject. As better understanding of this syndrome develops with neuroimaging and other advances, the classification of

TABLE **6.1.** Localization of apraxia

Type	Lesion(s)
Limb kinetic	Contralateral premotor area
Ideomotor	
Buccofacial	Left frontal operculum
Limb	Left perisylvian area or anterior corpus callosum
Axial	Left temporal lobe
Ideational	Left parietal lobe or diffuse involvement

apraxia may improve, but Liepmann's system provides a solid foundation. Table 6.1 lists these varieties and their putative neuroanatomic bases.

LIMB KINETIC APRAXIA

This variety of apraxia is the least frequently diagnosed, reflecting an uncertainty on whether it differs from mild paresis. Also known as melokinetic or innervatory apraxia, *limb kinetic apraxia* occurs in patients with limited lesions of the premotor area (the lateral portion of Brodmann area 6) or subjacent white matter, and it is a unilateral disorder opposite to the involved hemisphere (Hier, Gorelick, and Shindler 1987). These patients display a loss of the usual agility, efficiency, and precision of the affected side, most notably in the hand. Thus there may be difficulty with such tasks as picking up a coin from a table and buttoning a shirt. Although gross movements are normal and there is no incoordination, tasks such as pantomime and object manipulation present obstacles. Gestures used for pantomime appear clumsy, and playing cards may pose special difficulty. This is a syndrome in which motor acts requiring the most delicately organized cortical circuitry are compromised. Although more subtle than other motor disturbances, the deficit may temper what is otherwise considered a good recovery from a frontal lobe insult.

Limb kinetic apraxia has been a controversial entity, some authorities maintaining that the syndrome only reflects primary motor dysfunction (Strub and Black 1993). Although it is possible that limb kinetic apraxia reflects corticospinal dysfunction alone, it is noteworthy that pyramidal lesions in the internal capsule do not produce the syndrome (Hier, Gorelick, and Shindler 1987). A premotor cortical lesion may therefore cause a qualitatively distinct form of motor impairment.

IDEOMOTOR APRAXIA

Ideomotor apraxia is the most common of the apraxias and has been subjected to the most detailed analysis. The defining feature of this disorder is the failure of a patient to carry out a motor act to verbal command, even though comprehension of the request is preserved and the primary motor system for its execution is intact (Heilman and Rothi 2003). Thus there has been presumed to be a separation between the idea of an act and its performance. The syndrome may be evident in the oral (buccofacial), limb, or axial musculature. Patients fail most notably on pantomime tasks and show less severe impairment on imitation of the examiner or use of an actual object. In contrast to limb kinetic apraxia, ideomotor apraxia is bilateral (except when the right limbs are too paretic to be regarded as apraxic); in contrast to ideational apraxia (see below), ideomotor apraxia refers to a single action, not a series of movements.

Ideomotor praxis is tested by directing the patient to perform certain learned movements that require the oral, limb, and axial musculature. Examples would be "Blow out a match," "Flip a coin," "Swing a baseball bat," and "Comb your hair." Intransitive gestures can also be observed, such as those elicited by asking a patient to "cough," "wave good-bye," and "stand up." First the examiner asks for the action without demonstrating it; if the patient fails this task, imitation of the examiner's movement can sometimes improve performance. If still there is no success, some patients can improve on a transitive task by actually using the object involved (e.g., blowing out a lighted match). Inability to use the actual object indicates the most severe form of ideomotor apraxia. By advancing through this sequence, the examiner can develop a detailed understanding of the depth and nature of the apraxic disturbance.

Errors in ideomotor praxis can be made in several ways (Heilman and Rothi 2003). Most unmistakable is simply the failure to generate any response at all. Perseveration on a task performed previously is common (e.g., after successfully showing how to blow out a match, the patient then repeats that procedure when asked to suck on a straw). Sometimes a vocalization is produced instead of the desired action itself (e.g., saying "cough" rather than coughing). Finally, the use of a body part as an object is frequently seen (e.g., using the hand instead of the comb).

The localization of ideomotor apraxia has important implications. To begin, a detailed meta-analysis of patients with this syndrome concluded that, in the vast majority of cases, lesions occur in a distributed network of neocortical and white matter regions of the left hemisphere, and the basal ganglia are not involved (Pramstaller and Marsden 1996). The identification of a distributed network for praxis is consistent with the notion of ideomotor apraxia as a cerebral disconnection syndrome, first posited by Liepmann and then defended by Geschwind (1965). Thus there has come to be a Liepmann-Geschwind model of ideomotor apraxia that predicts the localization of lesions based on disconnection of critical cerebral areas (Figure 6.1).

The proposed model considers ideomotor apraxia to occur after damage to one or more of three different areas (Absher and Benson 1993). First, a lesion in

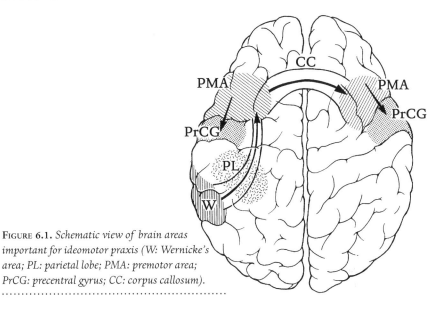

FIGURE **6.1.** *Schematic view of brain areas important for ideomotor praxis (W: Wernicke's area; PL: parietal lobe; PMA: premotor area; PrCG: precentral gyrus; CC: corpus callosum).*

the left parietal lobe can cause damage to the left arcuate fasciculus, interrupting the flow of verbal information anteriorly and preventing the motor system from receiving the direction to act (Benson et al. 1973). Conduction aphasia is commonly present. Second, a large lesion in the left premotor area can cause the deficit by interfering with the motor execution of the act, usually in association with nonfluent aphasia and right hemiparesis (Geschwind 1975). When there is hemiparesis, apraxia in the non-paretic left hand is sometimes called *sympathetic apraxia*. Finally, damage to the anterior corpus callosum can lead to ideomotor apraxia, evident only in the left hand (*callosal apraxia*; Geschwind and Kaplan 1962); the right hand is unaffected and no aphasia is present.

The disconnection theory of ideomotor apraxia, however, has not gone unchallenged. An issue of debate in this area centers on the possibility of left hemisphere cortical repositories for movement patterns. Liepmann had in fact originally suggested that the left hemisphere contained engrams responsible for skilled movements. These engrams for praxis might be stored in the frontal lobe, where programming for motor execution might take place (Devinsky 1992). Alternatively, others have made an argument for the existence of "visuo-kinesthetic" engrams or "praxicons" in the parietal lobe, pointing out that some apraxic patients can neither perform accurate movements nor recognize them in others and, therefore, must have lost the internal representation of the movements (Heilman, Rothi, and Valenstein 1982; Heilman and Rothi 2003). The debate between the disconnection and the praxicon theories has not been resolved, although it is not inconceivable that each may be correct depending on the patient and the lesion involved.

It should be apparent that ideomotor apraxia can occur with lesions virtually anywhere in the perisylvian language area of the left hemisphere, and indeed clinical experience bears out this prediction (Alexander et al. 1992). The low frequency of apraxia with right hemisphere lesions also suggests a dominance of the left hemisphere for learned movements not unlike its dominance for language (DeRenzi, Motti, and Nichelli 1980; Alexander et al. 1992). Left hemisphere dominance for praxis is also supported by studies of callosal apraxia (Graff-Radford, Welsh, and Godersky 1987). A notable finding has been that buccofacial apraxia seems to relate particularly to left frontal operculum and paraventricular white matter lesions (Tognola and Vignolo 1980; Alexander et al. 1992). Limb apraxia, on the other hand, has no specific perisylvian area with which it is associated (Alexander et al. 1992; Basso, Luzzatti, and Spinnler 1980; Kertesz and Ferro 1984; Basso et al. 1987). In the case of axial apraxia, perisylvian lesions do not appear to be involved, and the cerebral basis of axial praxis is unresolved; Geschwind (1975) suggested that descending pathways from the left temporal lobe to the pons and cerebellum could be responsible.

More recent studies have proposed the existence of *conceptual apraxia*, which is similar to ideomotor apraxia but differs in that the errors made by affected patients are related to the concept of the act and not its production (Heilman and Rothi 2003). Conceptual apraxia implies that content errors are made in contrast to the production errors of ideomotor apraxia. Thus, for example, patients with conceptual apraxia may fail to use a tool correctly because its function is not understood, whereas the ideomotor apraxic fails the task by not producing a movement that is well understood conceptually. The localization of lesions producing conceptual apraxia is unclear, although conceptual apraxia has been found to be common in patients with Alzheimer's disease (AD), in whom parietal damage can be assumed (Ochipa, Rothi, and Heilman 1992). The left parietal lobe is tentatively identified as the most likely site, and indeed conceptual and ideomotor apraxias are often reported to occur together (Heilman and Rothi 2003). However, the clinical utility of detecting conceptual apraxia is uncertain, and ideomotor apraxia remains the more widely recognized variety.

A related matter concerns the relationship of praxis to skills, often discussed in the context of procedural memory (Chapter 4). Although the neurologic literature on memory rarely intersects with that devoted to praxis, it may well be that skill learning and the acquisition of praxis are closely related or even identical phenomena. Whereas it has long been known that procedural memory does not require the medial temporal system (Chapter 4), its neuroanatomic basis has not been entirely clear. More recent research, however, has clarified this area, and there is mounting evidence that motor structures including the basal ganglia

(Martone et al. 1984; Lafosse et al. 2007) and the cerebellum (Sanes, Dimitrov, and Hallett 1990) participate in the acquisition of new motor skills. With practice and repetition, these motor skills may be increasingly incorporated into cortical areas and become gradually less dependent on subcortical structures. In a study bearing upon this hypothesis, a group of AD patients who had apraxia were found to have normal procedural learning as tested by a rotary pursuit task (Jacobs et al. 1999). This finding implies that what is initially learned as a skill in the basal ganglia and cerebellum eventually becomes encoded as a memory for praxis in the neocortex. It is reasonable to suppose that procedural learning is accomplished by subcortical gray matter structures, and then the motor memories so acquired are stored as cortical representations in the left hemisphere; praxis can therefore be regarded as the motor equivalent of remote declarative memory. Ideomotor apraxia in AD, in turn, would result from loss of learned motor skills as a result of cortical degeneration, while skill learning is better preserved because the subcortical gray matter regions by which it is accomplished remain relatively intact. This formulation also agrees with the observation that ideomotor apraxia is strongly associated with left hemisphere lesions sparing the basal ganglia (Pramstaller and Marsden 1996). Further study of the relationship between procedural learning and praxis clearly seems warranted.

The utility of detecting ideomotor apraxia lies first in its localizing value and also in the propensity of the syndrome to interfere with functional recovery in patients with cerebral disease. An association of this form of apraxia with focal left hemisphere lesions or, less often, the anterior corpus callosum has been established, and neuroimaging techniques are likely to further refine the neuroanatomy of the syndrome. In the rehabilitation of persons with neurologic disease, the deleterious effects of ideomotor apraxia can be disabling (Hanna-Pladdy, Heilman, and Foundas 2003). Patients with right hemiplegia who must use their left side, for example, may discover that acquisition of motor skills by the left hand is severely limited by sympathetic apraxia. Similarly, nonfluent aphasics with buccofacial apraxia may have difficulties producing speech sounds, which complicate the language problems that already exist and impede the rehabilitative process.

IDEATIONAL APRAXIA

The final apraxic category initially proposed by Liepmann is *ideational apraxia*, which in his view was the failure to perform a sequential motor act even though each constituent act could be performed in isolation (Liepmann 1920). Thus patients fail at a complex task because of a faulty overall plan, despite the preservation of individual acts (Lehmkuhl and Poeck 1981). Ideational apraxia is present, for example, when a patient cannot complete a series of actions such as folding a letter, inserting it in an envelope and sealing it, and applying a postage stamp, despite being able to perform each act by itself. Although damage to the left parietal lobe has been implicated in the genesis of ideational apraxia (Hier, Gorelick, and Shindler 1987; Devinsky 1992), the localization is unsettled since patients with diffuse cortical involvement from degenerative dementia also display the syndrome (Mendez and Cummings 2003). It is possible that parietal involvement in AD (Chapter 12) explains the failure to perform sequential acts, but alternatively, this deficit could imply frontal lobe dysfunction since it resembles the executive function deficits seen with bilateral prefrontal damage.

Unfortunately, other definitions of ideational apraxia have been advanced, and serious confusion exists in this area. The term has been employed to mean a deficit in the manipulation of actual objects or tool use (DeRenzi, Pieczuro, and Vignolo 1968), and the loss of knowledge of how to use an object or tool has been emphasized (DeRenzi and Lucchelli 1988). Others have interpreted this syndrome as a conceptual defect (Ochipa, Rothi, and Heilman 1989). None of these formulations of ideational apraxia, however, appears to indicate cerebral pathology with any precision, and its practical utility as a concept is limited.

REFERENCES

Absher, J. R., and Benson, D. F. Disconnection syndromes: an overview of Geschwind's contributions. *Neurology* 1993; 43: 862–867.

Alexander, M. P., Baker, E., Naeser, M. A., et al. Neuropsychological and neuroanatomical dimensions of ideomotor apraxia. *Brain* 1992; 115: 87–107.

Basso, A., Capitani, E., Della Sala, S., et al. Recovery from ideomotor apraxia: a study on acute stroke patients. *Brain* 1987; 110: 747–760.

Basso, A., Luzzatti, C., and Spinnler, H. Is ideomotor apraxia the outcome of damage to well-defined regions of the left hemisphere? Neuropsychological study of CAT correlation. *J Neurol Neurosurg Psychiatry* 1980; 43: 118–126.

Benson, D. F., Sheremata, W. A., Buchard, R., et al. Conduction aphasia. *Arch Neurol* 1973; 28: 339–346.

Denny-Brown, D. D. The nature of apraxia. *J Nerv Ment Dis* 1958; 126: 9–33.

DeRenzi, E., and Lucchelli, F. Ideational apraxia. *Brain* 1988; 111: 1173–1185.

DeRenzi, E., Motti, F., and Nichelli, P. Imitating gestures: a quantitative approach to ideomotor apraxia. *Arch Neurol* 1980; 37: 6–10.

DeRenzi, E., Pieczuro, A., and Vignolo, L. A. Ideational apraxia: a quantitative study. *Neuropsychologia* 1968; 6: 41–52.

Devinsky, O. *Behavioral Neurology: 100 Maxims.* St. Louis: Mosby Year Book; 1992.

Geschwind, N. The apraxias: neural mechanisms of disorders of learned movement. *Am Sci* 1975; 63: 188–195.

———. Disconnexion syndromes in animals and man. *Brain* 1965; 88: 237–294, 585–644.

Geschwind, N., and Kaplan, E. A human cerebral deconnection syndrome. *Neurology* 1962; 12: 675–685.

Graff-Radford, N. R., Welsh, K., and Godersky, J. Callosal apraxia. *Neurology* 1987; 37: 100–105.

Greene, J.D.W. Apraxia, agnosias, and higher visual function abnormalities. *J Neurol Neurosurg Psychiatry* 2005; 76 (Suppl V): v25–v34.

Hanna-Pladdy, B., Heilman, K. M., and Foundas, A. Ecological implications of ideomotor apraxia: evidence from physical activities of daily living. *Neurology* 2003; 60: 487–490.

Heilman, K. M., and Rothi, L.J.G. Apraxia. In: Heilman, K. M., and Valenstein, E., eds. *Clinical Neuropsychology*. 4th ed. New York: Oxford University Press; 2003: 215–235.

Heilman, K. M., Rothi, L. J., and Valenstein, E. Two forms of ideomotor apraxia. *Neurology* 1982; 32: 342–346.

Hier, D. B, Gorelick, P. B., and Shindler, A. G. *Topics in Behavioral Neurology and Neuropsychology*. Boston: Butterworths; 1987.

Jacobs, D. H., Adair, J. C., Williamson, D. J., et al. Apraxia and motor-skill acquisition in Alzheimer's disease are dissociable. *Neuropsychologia* 1999; 37: 875–880.

Kertesz, A., and Ferro, J. M. Lesion size and location in ideomotor apraxia. *Brain* 1984; 107: 921–933.

Kirshner, H. *Behavioral Neurology: Practical Science of Mind and Brain*. 2nd ed. Boston: Butterworth-Heinemann; 2002.

Lafosse, J. M., Corboy, J. R., Leehey, M. A., et al. MS vs. HD: can white matter and subcortical gray matter pathology be distinguished neuropsychologically? *J Clin Exp Neuropsychol* 2007; 29: 142–154.

Lehmkuhl, G., and Poeck, K. A disturbance in the conceptual organization of actions in patients with ideational apraxia. *Cortex* 1981; 17: 153–158.

Lewis, J. W. Cortical networks related to human use of tools. *Neuroscientist* 2006; 12: 211–231.

Liepmann, H. Apraxie. *Erbgn der Ges Med* 1920; 1: 516–543.

———. Das Krankheitsbild der Apraxie (motorischen Asymbolie auf Grund eines Falles von einseitiger Apraxie). *Monatsschrift für Psychiat Neurol* 1900; 8: 15–44, 102–132, 182–197.

———. Der weitere Kranksheitverlauf bei den einseitig Apraktischen und der Gehirnbefund auf Grund von Schnittserien. *Monatsschrift für Psychiat Neurol* 1906; 17: 217–243, 289–311.

———. *Drei Aufsätze aus dem Apraxiegebiet*. Berlin: Karger; 1908.

Liepmann, H., and Maas, O. Fall von linksseitiger Agraphie und Apraxie bei rechtsseitiger Lahmung. *J für Psychol Neurol* 1907; 10: 214–217.

Martone, M., Butters, N., Payne, M., et al. Dissociations between skill learning and verbal recognition in amnesia and dementia. *Arch Neurol* 1984; 41: 965–970.

Mendez, M. F., and Cummings, J. L. *Dementia: A Clinical Approach*. 3rd ed. Philadelphia: Butterworth-Heinemann; 2003.

Ochipa, C., Rothi, L.J.G., and Heilman, K. M. Conceptual apraxia in Alzheimer's disease. *Brain* 1992; 115: 1061–1071.

———. Ideational apraxia: a deficit in tool selection and use. *Ann Neurol* 1989; 25: 190–193.

Pramstaller, P. P., and Marsden, C. D. The basal ganglia and apraxia. *Brain* 1996; 119: 319–340.

Sanes, J. N., Dimitrov, B., and Hallett, M. Motor learning in patients with cerebellar dysfunction. *Brain* 1990; 113: 103–120.

Strub, R. L., and Black, F. W. *The Mental Status Examination in Neurology*. 3rd ed. Philadelphia: F. A. Davis; 1993.

Tognola, G., and Vignolo, L. A. Brain lesions associated with oral apraxia in stroke patients: a clinico-neuroradiological investigation with the CT scan. *Neuropsychologia* 1980; 18: 257–272.

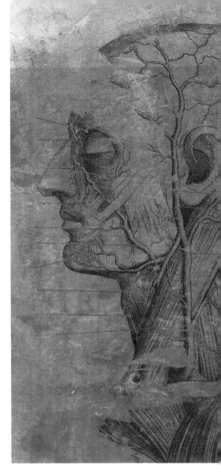

CHAPTER SEVEN

AGNOSIA

Agnosia is fundamentally a disorder of recognition (Bauer and Demery 2003; Greene 2005). Like aphasia and apraxia, it may follow a focal cerebrovascular event but can also be seen in degenerative disease. A patient with agnosia fails to recognize an object even when primary sensory modalities have registered its features adequately. From the Greek meaning "absence of knowledge," the term was first introduced by Sigmund Freud in his early monograph on aphasia (Freud 1891). Since then, no syndrome in behavioral neurology has engendered more debate, confusion, and controversy. It has been commented, for example, that the number of suggested mechanisms for visual agnosia nearly equals the number of reported cases of the syndrome (Benson and Greenberg 1969). A consideration of agnosia challenges the student of behavior because of the complexity of patients with various agnosias and the conceptual difficulties inherent in the organization of higher sensory function.

The detection of agnosia in an individual patient is rarely straightforward. Affected persons may not offer specific complaints that suggest agnosia, and findings may be subtle on examination. A visual agnosic, for example, often struggles to express visual dysfunction, reverting to generic complaints such as blurred vision or difficulty focusing that prompt an unrevealing optometric or ophthalmoscopic examination. Similarly, a patient with auditory agnosia commonly complains of hearing loss that results in an audiogram that will not explain the auditory symptoms. Recognizing agnosia from the clinical history requires special attention to potential higher sensory dysfunction, and when suspected, this can become a focus of the mental status examination. Detailed testing for agnosia is not necessary except in unusual cases, as the syndrome is sufficiently uncommon that cases can be considered on an individual basis (Chapter 2). Agnosic deficits may appear, however, in the course of the clinical history-taking; examples would be reports of difficulty recognizing a familiar object, face, or sound. Problems may also be uncovered in the mental status examination, such as comprehension deficit suggesting pure word deafness, and even in the elemental neurologic examination when a person has impairments in visual or tactile function or of higher sensory processing. Extra time and inge-nuity are often required to elicit and interpret the agnosias, but the results of this effort can be informative and clinically useful (Strub and Black 1993; Cummings and Mega 2003).

A standard definition of agnosia holds that it is "a normal percept that has somehow been stripped of its meaning" (Teuber 1968). This concept means that information from the external world is perceived but not recognized, so that its previously acquired meaning is no longer attached to it. A key feature of agnosia is that it exists only in a single sensory modality; an object that cannot be identi-fied through one modality can be recognized in another. A visual agnosic, for example, cannot identify a set of keys by sight but can easily do so when allowed to hear them jingle or feel them manually. Similarly, a word-deaf patient cannot understand spoken language but is easily able to read. Thus agnosia is modality specific and represents a disruption of a single sensory input system that associ-ates meaning with sensation. Implicit in this formulation is that agnosia differs from anomia; inability to name an object implies a loss of its word represen-tation and does not differ depending on the sensory system involved, whereas inability to recognize an object implies a loss of access to the encoded mean-ing that is normally activated through a single sensory modality. Moreover, an anomic patient recognizes the meaning of an object that cannot be named.

In clinical practice, demonstration of agnosia requires the absence of pri-mary sensory loss, acute confusional state, amnesia, and aphasia, all of which can mimic the syndrome. Because agnosia generally occurs in the context of large hemispheric lesions, this requirement is frequently difficult to meet, and unequivocal cases are rare in the literature. Indeed, vigorous opposition to the concept of agnosia, especially visual agnosia, has been raised by critics who claim that agnosia merely reflects a combination of primary sensory loss and intellec-

TABLE 7.1. Localization of agnosia

Type	Lesion(s)
Visual	
Apperceptive	Bilateral occipital lobe
Associative	Bilateral occipitotemporal region
Object agnosia	Left or bilateral occipitotemporal region
Prosopagnosia	Right or bilateral occipitotemporal region
Central achromatopsia	Bilateral occipitotemporal region
Simultanagnosia	Bilateral occipitoparietal region
Akinetopsia	Bilateral occipitoparietal region
Auditory	
Pure word deafness	Left or bilateral temporal lobe
Auditory sound agnosia	Right or bilateral temporal lobe
Tactile	Contralateral parietal lobe

tual deterioration (Bay 1953; Bender and Feldman 1972). However, recent years have witnessed substantial progress in the identification of agnosias, and a better understanding of the neuroanatomic basis of these intriguing disorders is emerging (Bauer and Demery 2003; Greene 2005). Three general categories have been traditionally considered: visual, auditory, and tactile (Bauer and Demery 2003). Table 7.1 summarizes the presumed localization of these various syndromes.

VISUAL AGNOSIA

The first point to be considered is the issue of *cortical blindness*. In this syndrome, extensive bilateral destruction of the primary visual cortices in the occipital lobes (Brodmann area 17, also called the striate or calcarine cortex) or their subjacent white matter leads to blindness with preserved pupillary responses and absent optokinetic nystagmus (Symonds and Mackenzie 1957). The syndrome is often accompanied by what has been called denial of blindness, a form of anosognosia known as *Anton's syndrome* (Anton 1899). Patients affected by Anton's syndrome act as though their vision is intact, and their propensity to collide with walls and furniture may be explained away with complaints about poor lighting or the inadequacy of their glasses. The origin of this visual anosognosia is uncertain, but it may be that patients with Anton's syndrome do in fact "see" in a limited fashion—the syndrome has been noted to be similar to *blindsight*, in which a cortically blind patient has residual visual perception resulting from sparing of the superior colliculi, pulvinars, and parietal cortices (Damasio, Tranel, and Rizzo 2000).

Another point to be clarified is the visual agnosia reported as part of the Klüver-Bucy syndrome, a disorder first produced in monkeys by the surgical ablation of both anterior temporal lobes (Klüver and Bucy 1939). This syndrome, which is also encountered in several human conditions affecting the temporal lobes (Lilly et al. 1983), does include among its manifestations a disturbance of vision, but the deficit is essentially one of failure to recognize the emotional significance of a visually presented object. For example, monkeys so affected fail to distinguish edible from inedible objects. The alternate term *psychic blindness* (Klüver and Bucy 1939), then, may be preferable to visual agnosia, and the disorder more appropriately considered a visual-limbic disconnection than a disturbance of higher visual function. Visual agnosia is properly considered a cognitive disorder, whereas the Klüver-Bucy syndrome involves a disturbance of emotional processing. Chapter 9 will take up the Klüver-Bucy syndrome in more detail.

The terminology surrounding the topic of *visual agnosia* and its variants is difficult, and no classification system has gained universal acceptance. The problem of determining a taxonomy for the visual agnosias is an important one for it influences how theories of higher visual function are developed (Farah 1990). Despite the uncertainty in this field, a basic distinction that enjoys widespread popularity is that proposed by Lissauer into apperceptive and associative varieties (Lissauer 1890). This division presupposes a two-step process of visual recognition: the synthesis of visual elements into a unified image ("apperception") and the matching of the image with previously encoded visual information ("association"). Although this distinction may not be straightforward in individual cases (DeRenzi and Lucchelli 1993), we will employ Lissauer's terminology as an organizing framework.

Apperceptive visual agnosia, the more common of the two varieties, is a failure to recognize objects because of a failure to perceive them. These patients have bilateral, albeit not complete, occipital lobe lesions and areas of impaired elemental vision, but they fail to recognize objects even in preserved areas of their visual fields. This syndrome differs from cortical blindness by the less extensive occipital destruction that allows normal visual acuity in spared fields, but forms and shapes are not recognized, and affected patients are unable to draw misidentified items or match them to samples (Benson and Greenberg 1969). The specific deficits beyond these general features have varied greatly in the reported cases (Bauer and Demery 2003).

Associative visual agnosia, less common but more convincing when it is encountered, involves a defect in recognizing a well-perceived visual stimulus. Again there is impairment of visual recognition, but these patients are clearly not blind in a functional sense; unlike apperceptive visual agnosics, they can make drawings of pictures they cannot recognize or name, and they can match drawings and pictures to samples (Rubens and Benson 1971). In addition, deficits in color naming and pure alexia (Chapter 5) are frequently seen in associative visual agnosia, leading Geschwind to explain the syndrome as a visual-verbal disconnection (Geschwind 1965). However, there have been autopsy-verified cases of associative visual agnosia with bilateral lesions of the occipitotemporal cortex and subjacent white matter (Benson, Segarra, and Albert 1974; Albert et al. 1979), suggesting destruction of the areas retaining memories for visual objects as an alternative explanation.

Since Lissauer's original apperceptive versus associative distinction, other more specific categories of visual agnosia have been examined. Each of these entities—object agnosia, prosopagnosia, central achromatopsia, and simultanagnosia—identifies a breakdown in some aspect of visual recognition. For our purposes, these will all be considered variants of visual associative agnosia since primary visual cortices are typically intact while some portions of the visual association areas are damaged (Table 7.1). The neuroanatomic basis of these disorders is becoming clarified through case studies and functional neuroimaging techniques, and the study of higher visual function is an area of increasingly active research.

Object agnosia refers specifically to difficulty recognizing objects such as a cup, chair, or clock. Object recognition has been thought to depend on either left unilateral or bilateral occipitotemporal regions (lingual, fusiform, and parahippocampal gyri), and infarction in the cortex and underlying white matter of these areas has caused object agnosia (Bauer and Demery 2003). The term *optic aphasia* refers to a milder form of this disorder in which visually presented objects can be recognized but not named, and auditory and tactile naming are normal (Bauer and Demery 2003). Positron emission tomography (PET) data from normal subjects have indicated that object recognition may depend predominantly on left occipitotemporal regions (Sergent, Ohta, and Macdonald 1992). While the issue of lateralization in this syndrome has not been resolved, the frequent occurrence of right homonymous hemianopia and alexia with object agnosia argues for a left-sided advantage for object recognition.

In contrast, facial recognition may depend on right hemisphere structures more than left. The syndrome of *prosopagnosia*, or face agnosia, involves selective impairment of the recognition of faces, which may even extend to the inability to recognize one's own face in a mirror (Bauer and Demery 2003). Some debate has revolved around whether bilateral lesions are necessary for the development of prosopagnosia. Some cases have been found to have bilateral lesions of the occipitotemporal cortex and white matter (Damasio, Damasio, and Van Hoesen 1982), but other observations have indicated that a single right occipitotemporal lesion is sufficient to cause the syndrome (Michel, Poncet, and Signoret 1989). Although both hemispheres participate in face recognition, the right visual association cortices appear to have an advantage in this capacity (Damasio, Tranel, and Damasio 1990). PET studies in normal subjects have also been consistent with this notion (Sergent, Ohta, and Macdonald 1992). Thus face and object processing appear to be dissociated so that the right and left hemispheres are functionally specialized for these respective tasks.

Central achromatopsia is a selective loss of color vision due to occipitotemporal damage (Damasio et al. 1980). When both hemispheres are involved, the achromatopsia is complete, and unilateral lesions can cause quadrantic or hemifield color deficits. Patients complain of a "gray" or "washed-out" look in affected areas of vision. Neurophysiological studies of laboratory animals and functional neuroimaging in humans have documented the existence of color-coded neurons in the fusiform and lingual gyri that are dedicated to central color processing (Damasio, Tranel, and Rizzo 2000). A different syndrome is *color anomia*, often occurring with pure alexia (Geschwind and Fusillo 1966), in which color naming but not color recognition is impaired. Both of these must be distinguished from inherited color blindness and acquired disorders of the optic nerves such as multiple sclerosis that compromise color vision.

The intriguing problem of *simultanagnosia* (simultaneous agnosia) represents a disorder involving failure to synthesize all the elements of a picture or scene, even though its constituent components can be recognized in isolation (Wolpert 1924). Simultanagnosia is one of the three features of *Balint's syndrome* (Balint 1909), a syndrome caused by bilateral occipitoparietal lesions; the other components are so-called *ocular apraxia* (oculomotor apraxia or psychic paralysis of gaze), an inability to shift gaze voluntarily from a fixation point, and *optic ataxia*, an impairment of visually guided movements due to a defect in stereopsis (depth perception). An often-reported cause of Balint's syndrome has been stroke from hypotension (Damasio, Tranel, and Rizzo 2000). More recently, cases of simultanagnosia heralding the onset of degenerative dementia have been presented, and in some patients an atypical posterior distribution of changes consistent with Alzheimer's disease (AD) has been found (Graff-Radford et al. 1993). The complete Balint's syndrome has also been described recently in cases of posterior cortical atrophy (Benson, Davis, and Snyder 1988), most of which will be found at autopsy to have AD. The suggestion has been advanced that simultanagnosia is a disorder of vigilance in which attentional mechanisms of both posterior hemispheres are disturbed and there is difficulty maintaining conscious experience of visual stimuli—the patient therefore "looks but does not see" (Rizzo and Hurtig 1987).

Finally, rare cases of *akinetopsia* have been described in which affected patients have selective difficulty perceiving motion. Also known as motion blindness, this syndrome has been observed in neurodegenerative disease and traumatic brain injury, and bilateral posterior cortical dysfunction has been implicated (Rizzo,

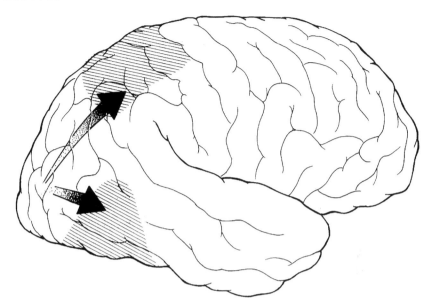

FIGURE 7.1. *Visual association areas in the parietal and temporal lobes, responsible for processing spatial and object properties of the visual image.*

Nawrot, and Zihl 1995; Pelak and Hoyt 2005; Tsai and Mendez 2009). Specific damage to the occipitoparietal cortices appears to be the most probable etiology of akinetopsia (Rizzo, Nawrot, and Zihl 1995; Tsai and Mendez 2009).

The anatomy of higher visual processing has become clarified from studies in higher primates and normal humans, and this information helps illuminate the cerebral basis of the visual agnosias (Figure 7.1). Much evidence supports the presence of two parallel systems, *ventral* and *dorsal*, that serve to process primary visual information once it is assembled by the occipital cortices (Mishkin, Ungerleider, and Macko 1983; Haxby et al. 1991). These systems have also been labeled *parvocellular* and *magnocellular*, respectively, corresponding to separate populations of neurons in the lateral geniculate nucleus of the thalamus that provide visual input to the cortical processing areas. The ventral system, involving the temporal lobe (Brodmann area 37), deals with object properties such as shape, form, and color, and the dorsal system, extending into the parietal lobe (Brodmann area 7), with spatial properties such as location, motion, and stereopsis. More succinctly, these two parallel neural streams may also be considered the *what* and the *where* systems.

In light of this dichotomy, all the variants of visual agnosia we have considered can be reconciled. Apperceptive visual agnosia is related to bilateral occipital damage that prevents the adequate perception of visual stimuli. Associative visual agnosia is a term that best serves as a generic designation for recognition deficits that affect various portions of visual association areas. The ventral system, responsible for object properties, is disrupted in object agnosia, prosopagnosia, and central achromatopsia, whereas simultanagnosia and akinetopsia occur when the dorsal system, responsible for spatial properties, is damaged. After a century of perplexing case reports and cryptic theorizing, this dorsal versus ventral stream formulation offers a systematic framework for visual agnosia that should usefully guide further study of this fascinating topic.

Before leaving the visual agnosias, brief mention of defects in *visual imagery* should be made. Imagery means "seeing in the mind's eye," and, in contrast to similar phenomena that characterize visual hallucinations and dreaming, it is a volitional mental act. Activation of stored information in the visual system is required for visual imagery, and imagery defects probably reflect damage to the same visual association areas where knowledge of the external environment is represented (Farah 1990). Case studies have disclosed that visual imagery impairments may selectively implicate the ventral and dorsal systems, and these deficits parallel the recognition disorder associated with these regions. That is, loss of object-color imagery is associated with prosopagnosia and achromatopsia due to temporooccipital damage, and loss of visuospatial imagery occurs with visual disorientation from parietooccipital damage (Levine, Warach, and Farah 1985). Although it might be imagined that the right hemisphere excels at visual imagery, the left hemisphere also appears to participate, and clinical and experimental evidence favors the view that both hemispheres make important contributions to visual imagery (Kosslyn 1988; Sergent 1990).

AUDITORY AGNOSIA

Auditory agnosia is an impairment in the ability to recognize speech or nonverbal sounds in the presence of adequate hearing. This syndrome is similar to visual agnosia in that bilateral cerebral lesions are often implicated, but in this case the damage centers on the temporal lobes. Analogous with the problem of cortical blindness, extensive bilateral involvement results in *cortical deafness*; here the damage is in the primary auditory cortices (Heschl's gyri; Brodmann areas 41 and 42) or their subjacent white matter (Earnest, Monroe, and Yarnell 1977). The syndrome of auditory agnosia is less dramatic, and its diagnosis requires the demonstration of a modality-specific impairment in the recognition of auditory stimuli. The possibility of applying the distinction between Lissauer's apperceptive and associative visual agnosias to the auditory system has been suggested, but the clinical differentiation and putative neuroanatomy of these potential varieties have proven problematic (Mesulam 2000). Another distinction in the auditory agnosia literature, however, has proven useful and will guide our discussion: auditory word agnosia, or pure word deafness, and auditory sound agnosia.

Pure word deafness is a syndrome of impaired recognition of speech sounds. Voices are heard but words do not make sense. Because the ears are normal, hearing as tested by pure tone audiometry is intact. Although the syndrome of pure word deafness is closely allied to Wernicke's aphasia (Chapter 5) and may indeed evolve from this aphasia in a recovering patient, it is distinguished from aphasia by the preservation of spontaneous speech, reading, and writing (Coslett, Brashear, and Heilman 1984). Precise delineation of the two syndromes can be difficult, however, since word-deaf patients usually have repetition impairment and may have paraphasic speech. The pathologic anatomy of these cases is thought to involve a disconnection of the primary auditory area on both sides (Heschl's gyri, Brodmann areas 41 and 42) from Wernicke's area (Geschwind 1970), and lesions may be bilateral (Auerbach et al. 1982) or unilateral left temporal (Albert and Bear 1974). The rarity of pure word deafness relates in part to the discrete location of lesions necessary to disconnect Wernicke's area from auditory input without at the same time destroying it.

Auditory sound agnosia is a nonverbal counterpart of pure word deafness. It is a less common disorder, although conceivably its lower incidence results from its less obvious symptomatology, which rarely prompts medical consultation. Here the deficit involves recognition of nonverbal sounds, such as bells ringing and dogs barking, in the presence of normal hearing and intact language. Although the neuroanatomic basis of this syndrome is uncertain, it may be a right hemisphere analog of pure word deafness, since reported cases have documented either right-side (Spreen, Benton, and Fincham 1965) or bilateral (Albert et al. 1972) lesions that interrupt the input of nonverbal sounds to the right hemisphere auditory processing system. Studies of patients with unilateral right or left temporoparietal strokes have suggested that the cortical auditory areas of both hemispheres participate in the recognition of nonverbal sounds but that the right hemisphere is concerned with sound discrimination and the left hemisphere with semantic processing of the sound (Schnider et al. 1994).

TACTILE AGNOSIA

Tactile agnosia is an impairment of the ability to recognize objects tactually when peripheral sensory modalities of pain, temperature, light touch, vibration, and proprioception are intact. This syndrome is a unilateral disorder of higher sensory function that can be found in the hand contralateral to the side of a parietal lobe lesion, reflecting the neuroanatomic organization of the somatosensory system. As is true for all the agnosias, acute confusional state, amnesia, and aphasia should also be excluded before a diagnosis of tactile agnosia can be made, and the patient must be shown to have a modality-specific deficit in tactile sensation. Testing for tactile agnosia is accomplished by asking a patient to identify objects applied to one hand, such as a coin, key, or paper clip, without the use of visual or auditory cues.

Tactile agnosia is equivalent to *astereognosis*, a disorder of "cortical" sensation frequently sought in the routine neurologic examination. Astereognosis is a deficit in tactile recognition typically resulting from a lesion in the contralateral primary sensory cortex of the parietal lobe. Together with the finding of *agraphesthesia*, impaired recognition of numbers or letters traced on the skin of the palm, astereognosis has considerable utility in the detection of contralateral parietal lobe lesions.

The precise localization of parietal lobe lesions causing tactile agnosia has not been fully settled (Caselli 1991; Bauer and Demery 2003). Studies of tactile recognition have found clinical deficits to correlate with lesions in the hand region of the contralateral postcentral gyrus (Brodmann areas 3, 1, and 2; Corkin, Milner, and Rasmussen 1970; Roland and Larsen 1976), and tactile agnosia has also been documented in a patient with a more posterior lesion in Brodmann areas 40 and 39 (Reed and Caselli 1994). It should also be mentioned that tactile agnosia has been reported in the context of impaired spatial orientation seen in right hemisphere lesions (Semmes 1965), and that occasional cases of left-hand tactile agnosia may represent a modality-specific naming deficit due to callosal disconnection (Geschwind and Kaplan 1962).

REFERENCES

Albert, M. L., and Bear, D. Time to understand: a case study of word deafness with reference to the role of time in auditory comprehension. *Brain* 1974; 97: 373–384.

Albert, M. L., Soffer, D., Silverberg, R., and Reches, A. The anatomic basis of visual agnosia. *Neurology* 1979; 29: 876–879.

Albert, M. L., Sparks, R., von Stockert, T., and Sax, D. A case of auditory agnosia: linguistic and nonlinguistic processing. *Cortex* 1972; 8: 427–443.

Anton, G. Über die Selbstwahrnehmungen der Herderkrankungen des Gehirns durch den Kranken bei Rindenblindheit und Rindentaubheit. *Arch für Psychiatr* 1899; 11: 227–229.

Auerbach, S. H., Allard, T., Naeser, M., et al. Pure word deafness: analysis of a case with bilateral lesions and a defect at the prephonemic level. *Brain* 1982; 105: 271–300.

Balint, R. Seelenlähmung des "Schauens," optische Ataxie, raumliche Störung der Aufmerksamkeit. *Monatsschrift fur Psychiat Neurol* 1909; 25: 51–81.

Bauer, R. M, and Demery, J. A. Agnosia. In: Heilman, K. M., and Valenstein, E., eds. *Clinical Neuropsychology*. 4th ed. New York: Oxford University Press; 2003: 236–295.

Bay, E. Disturbances of visual perception and their examination. *Brain* 1953; 76: 515–550.

Bender, M. B., and Feldman, M. The so-called "visual agnosias." *Brain* 1972; 95: 173–176.

Benson, D. F., Davis, R. J., and Snyder, B. D. Posterior cortical atrophy. *Arch Neurol* 1988; 45: 789–793.

Benson, D. F., and Greenberg, J. P. Visual form agnosia. *Arch Neurol* 1969; 20: 82–89.

Benson, D. F., Segarra, J., and Albert, M. L. Visual agnosia-prosopagnosia: a clinicopathologic correlation. *Arch Neurol* 1974; 30: 307–310.

Caselli, R. J. Rediscovering tactile agnosia. *Mayo Clin Proc* 1991; 66: 129–142.

Corkin, S., Milner, B., and Rasmussen, T. Somatosensory thresholds: contrasting effects of postcentral-gyrus and posterior parietal lobe excision. *Arch Neurol* 1970; 23: 41–58.

Coslett, H. B., Brashear, H. R., and Heilman, K. M. Pure word deafness after bilateral primary auditory cortex infarcts. *Neurology* 1984; 34: 347–352.

Cummings, J. L., and Mega, M. S. *Neuropsychiatry and Behavioral Neuroscience*. New York: Oxford University Press; 2003.

Damasio, A. R., Damasio, H., and Van Hoesen, G. W. Prosopagnosia: anatomic basis and behavioral mechanisms. *Neurology* 1982: 32: 331–341.

Damasio, A. R., Tranel, D., and Damasio, H. Face agnosia and the neural substrates of memory. *Ann Rev Neurosci* 1990; 13: 89–109.

Damasio, A. R., Tranel, D., and Rizzo, M. Disorders of complex visual processing. In: Mesulam, M.-M., ed. *Principles of Behavioral and Cognitive Neurology*. 2nd ed. New York: Oxford University Press; 2000: 332–372.

Damasio, A. R., Yamada, T., Damasio, H., et al. Central achromatopsia: behavioral, anatomic, and physiologic aspects. *Neurology* 1980; 30: 1064–1071.

DeRenzi, E., and Lucchelli, F. The fuzzy boundaries of apperceptive agnosia. *Cortex* 1993; 29: 187–225.

Earnest, M. P., Monroe, P. A., and Yarnell, P. R. Cortical deafness: demonstration of the pathologic anatomy by CT scan. *Neurology* 1977; 27: 1172–1175.

Farah, M. J. *Visual Agnosia*. Cambridge: MIT Press; 1990.

Freud, S. *On Aphasia*. New York: International Universities Press; 1891.

Geschwind, N. Disconnexion syndromes in animals and man. *Brain* 1965; 88: 237–294, 585–644.

———. The organization of language and the brain. *Science* 1970; 170: 940–944.

Geschwind, N., and Fusillo, M. Color-naming deficits in association with alexia. *Arch Neurol* 1966; 15: 137–146.

Geschwind, N., and Kaplan, E. A human cerebral deconnection syndrome. *Neurology* 1962; 12: 675–685.

Graff-Radford, N. R., Bolling, J. P., Earnest, F., et al. Simultanagnosia as the initial sign of degenerative dementia. *Mayo Clin Proc* 1993; 68: 955–964.

Greene, J.D.W. Apraxia, agnosias, and higher visual function abnormalities. *J Neurol Neurosurg Psychiatry* 2005; 76 (Suppl V): v25–v34.

Haxby, J. V., Grady, C. L., Horwitz, B., et al. Dissociation of object and spatial visual processing pathways in human extrastriate cortex. *Proc Natl Acad Sci* 1991; 88: 1621–1625.

Klüver, H., and Bucy, P. C. Preliminary analysis of functions of the temporal lobes in monkeys. *Arch Neurol Psychiat* 1939; 42: 979–1000.

Kosslyn, S. M. Aspects of a cognitive neuroscience of mental imagery. *Science* 1988; 240: 1621–1626.

Levine, D. N., Warach, J., and Farah, M. Two visual systems in visual imagery: dissociation of "what" and "where" in imagery disorders due to bilateral posterior cerebral lesions. *Neurology* 1985; 35: 1010–1018.

Lilly, R., Cummings, J. L., Benson, D. F., and Frankel, M. The human Klüver-Bucy syndrome. *Neurology* 1983; 33: 1141–1145.

Lissauer, H. Ein Fall von Seelenblindheit nebst einen Beitrage zur Theorie derselben. *Arch für Psychiat* 1890; 21: 222–270.

Mesulam, M.-M. Behavioral neuroanatomy: large-scale networks, association cortex, frontal systems, the limbic system, and hemispheric specializations. In: Mesulam, M.-M. *Principles of Behavioral and Cognitive Neurology*. 2nd ed. New York: Oxford University Press; 2000: 1–120.

Michel, F., Poncet, M., and Signoret, J. L. Les Lésions responsables de la prosapagnosie sont-elles toujours bilatérales? *Rev Neurol* 1989; 146: 764–770.

Mishkin, M., Ungerleider, L. G., and Macko, K. A. Object vision and spatial vision: two cortical pathways. *Trends Neurosci* 1983; 6: 414–417.

Pelak, V. S., and Hoyt, W. F. Symptoms of akinetopsia associated with traumatic brain injury and Alzheimer's disease. *Neuro-Ophthalmology* 2005; 29: 137–142.

Reed, C. L., and Caselli, R. J. The nature of tactile agnosia: a case study. *Neuropsychologia* 1994; 32: 527–539.

Rizzo, M., and Hurtig, R. Looking but not seeing: attention, perception, and eye movements in simultanagnosia. *Neurology* 1987; 37: 1642–1648.

Rizzo, M., Nawrot, M., and Zihl, J. Motion and shape perception in cerebral akinetopsia. *Brain* 1995; 118: 1105–1127.

Roland, P. E., and Larsen, B. Focal increases of cerebral blood flow during stereognostic testing in man. *Arch Neurol* 1976; 33: 551–558.

Rubens, A. B., and Benson, D. F. Associative visual agnosia. *Arch Neurol* 1971; 24: 305–316.

Schnider, A., Benson, D. F., Alexander, D. N., and Schnider-Klaus, A. Non-verbal environmental sound recognition after unilateral hemispheric stroke. *Brain* 1994; 117: 281–287.

Semmes, J. A non-tactual factor in astereognosis. *Neuropsychologia* 1965; 3: 295–314.

Sergent, J. The neuropsychology of visual image generation: data, method, and theory. *Brain Cogn* 1990; 13: 98–129.

Sergent, J., Ohta, S., and Macdonald, B. Functional neuroanatomy of face and object processing. *Brain* 1992; 115: 15–36.

Spreen, O., Benton, A. L., and Fincham, R. W. Auditory agnosia without aphasia. *Arch Neurol* 1965; 13: 84–92.

Strub, R. L., and Black, F. W. *The Mental Status Examination in Neurology*. 3rd ed. Philadelphia: F. A. Davis; 1993.

Symonds, C., and Mackenzie, I. Bilateral loss of vision from cerebral infarction. *Brain* 1957; 80: 415–455.

Teuber, H. L. Alteration of perception and memory in man. In: Weiskrantz, L., ed. *Analysis of Behavioral Change*. New York: Harper and Row; 1968: 274–328.

Tsai, P.-H., and Mendez, M. F. Akinetopsia in the posterior cortical variant of Alzheimer disease. *Neurology* 2009; 73: 731–732.

Wolpert, I. Die Simultanagnosie: Störung der Gesamtauffassung. *Z Ges Neurol Psychiatr* 1924; 93: 397–415.

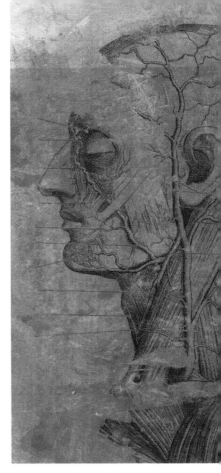

CHAPTER EIGHT

RIGHT HEMISPHERE
SYNDROMES

Lesions of the right hemisphere produce some of the most intriguing yet problematic syndromes in behavioral neurology. Despite a substantial amount of information that has appeared recently, these disorders present special difficulties since they typically do not reveal themselves through the usual route of verbal mediation. Right hemisphere disorders, although they may often be disabling in real-world settings, are often difficult for the patient and family or caregiver to describe and may escape unnoticed in the routine mental status examination unless specifically sought. When they are recognized, these disturbances can still be elusive because they implicate more intuitive, nonverbal cognitive systems for which our standard verbal accounts may be inadequate. Even the terminology for these syndromes is difficult because linguistic descriptions cannot necessarily capture inherently nonverbal phenomena. It is not by coincidence that the best-developed notions of brain-behavior relationships center on the language system, which is so much more accessible to the examiner (Chapter 5). One of the many challenges facing behavioral neurology is the continued development of a satisfactory lexicon and a meaningful taxonomy of right hemisphere functions and their dissolution.

Many of the syndromes discussed in this chapter represent a breakdown in the brain's ability to orient the body within external space and formulate an adequate behavioral repertoire within it. We live in a three-dimensional world within which we must attend to all relevant information, analyze and then reassemble it, remember important components, and generate appropri-

TABLE **8.1.** Right hemisphere syndromes

| Constructional apraxia |
| Neglect |
| Spatial disorientation |
| Dressing apraxia |
| Aprosody |
| Amusia |
| Emotional disorders |

ate motor responses. None of this can be adequately supported by the left hemisphere linguistic system, and a large body of evidence does indeed support the idea that the right hemisphere is specialized for these tasks. In light of the major importance of visuospatial integration—and the disabling syndromes that right hemisphere damage often produces—reference to this hemisphere as "minor" is unjustified. Even though our culture strongly rewards linguistic abilities, many other equally adaptive capacities are mediated by the right hemisphere, which itself can be considered dominant for a variety of functions. This chapter will attempt to illustrate the extent to which the right hemisphere contributes to the human mental life. Table 8.1 lists the major right hemisphere syndromes. The concluding section on the emotional aspects of right hemisphere function will form a transition to additional discussion of emotion and personality in Chapters 9 and 10.

CONSTRUCTIONAL APRAXIA

Constructional apraxia is a term widely used to refer to impairment in the production of drawings. People with this syndrome fail to correctly draw or copy three- or even two-dimensional shapes, and these deficits may translate into problems in everyday life such as difficulty with directions, getting lost while driving, and being unable to return safely home from a long walk. Although criticized for not being a true apraxia (Geschwind 1975), constructional apraxia has survived as the most commonly used designation; alternative terms—none entirely satisfactory—include visuospatial dysfunction, constructional disturbance, and apractagnosia (Cummings and Mega 2003). Avoiding the debate on whether this deficit qualifies as a true apraxia, we will defer to the more standard usage and describe the cerebral lesions most often responsible for impairments of this sort. Right hemisphere lesions are more often responsible for clinically significant syndromes, but it is worth remembering that left-side lesions also cause constructional disturbances that are qualitatively distinct from the deficits that follow right-side involvement.

A first point is that constructional apraxia usually implies a neurologic disease because the idiopathic psychiatric disorders do not ordinarily disrupt constructional ability (Cummings and Mega 2003). Further, it has been known for many decades that patients who struggle the most with producing two- and three-dimensional drawings are considerably more likely to have a right than a left hemisphere lesion. Within the right hemisphere, both parietal and frontal lesions can produce this syndrome, and it has been observed that frontal lesions disrupt drawing more than copying, whereas parietal lesions affect copying more than drawing (Cummings and Mega 2003). In particular, however, the right parietal lobe has emerged as most closely associated with constructional apraxia (Piercy, Hecaen, and de Ajuriaguerra 1960; Mesulam 1981). Drawing errors made by patients with right parietal lesions may typically include left hemineglect (as reviewed below), loss of perspective, and a tendency to work from right to left; these individuals may also display the phenomenon of "closing in," where the copy is placed too close to or even on top of the original (Hecaen and Albert 1978; Kaplan 1983; Cummings and Mega 2003). In contrast, patients with left parietal lesions may also be found to have constructional apraxia, although the impairment is milder and less persistent; their problems include simplification, omissions, and loss of internal detail (Hecaen and Albert 1978; Kaplan 1983; Cummings and Mega 2003).

NEGLECT

More than any other syndrome, the phenomenon of neglect dramatically illustrates the role of the right hemisphere in mediating visuospatial competence (Mesulam 1981; Heilman, Watson, and Valenstein 2003). *Neglect* is a breakdown in the ability to notice, report, or respond to stimuli, and when such stimuli presented contralateral to the side of a cerebral lesion are so ignored, *hemineglect* is the appropriate term (Mesulam 1981; Heilman, Watson, and Valenstein 2003; Greene 2005). Although mild forms of hemineglect can sometimes be seen after left hemisphere lesions, the syndrome is much more severe and frequent after right-side damage (Mesulam 1981; Denes et al. 1982). Patients with hemineglect display a remarkable array of behaviors, such as dressing only one side of the body, shaving just one side of the face, and eating only one side of a plate of food. These deficits are brought out more systematically by double simultaneous testing for sensory extinction, line-bisection and line-cancellation tasks, and drawing tests such as clock drawing with the hands at 11:10 (Cummings and Mega 2003). It is as though one half of the outer world is no longer available to these individuals, even though primary sensation is intact.

The often poor outcome of right hemisphere patients with hemineglect (Denes et al. 1982; Heilman, Watson, and Valenstein 2003) testifies to the importance of the right hemisphere in overall adaptation. When one half of one's extrapersonal space, and often one half of one's body as well, are ignored because of a brain lesion, significant hindrance to normal daily living can be expected. This deficit may be far more hazardous than loss of a sensory modality that also deprives a patient of one half his or her normal function. A hemianopia, for example, might seem to deprive a patient of one half the visual world, but in fact most people learn to adapt by simply turning the head as an adjustment; neglect patients, on the other hand, lack the capacity to attend to the portion of the world they are missing, and full vision is of little use in this situation. Patients with hemineglect are therefore at higher risk driving a motor vehicle, for example, than are hemianopic individuals.

Other deficits can be observed in patients with neglect. One problem is *anosodiaphoria* (Critchley 1953), which refers to unconcern about left hemiparesis or hemiplegia even as it is acknowledged by the patient. More serious is actual unawareness of left hemiplegia, a deficit known as *anosognosia* (Babinski 1914). A severely impaired patient may even express firm *denial* of the deficit, even to the point of denying that the affected limbs belong to him (Cummings and Mega 2003). Still another problem allied with neglect is *motor impersistence*, or failure to persist at a variety of willed acts, such as eye closure, arm extension, and tongue protrusion. Typically seen in left hemiparetic patients with neglect (Hier, Mondlock, and Caplan 1983; Kertesz et al. 1985), motor impersistence indicates a vigilance deficit that implies right frontal dysfunction. Again, it is apparent that recovery from a right hemisphere lesion producing these deficits might be difficult indeed.

The anatomy of neglect has been worked out in some detail (Figure 8.1). It seems clear that the right hemisphere is more involved in visuospatial surveillance than the left (Heilman and Van Den Abell 1980), and it has been nominated as the one dominant for attention (Weintraub and Mesulam 1987). Neglect seems to be particularly strongly correlated with destruction of the right parietal lobe (Mesulam 1981; Cummings and Mega 2003). It appears that the left hemisphere does attend to the right hemispace, and therefore right hemineglect can be a feature of acute left hemisphere lesions, but the right hemisphere attends to *both* sides of the environment. Thus an acute right hemisphere lesion can result in impressive left hemineglect since there is little ability of the left hemisphere to compensate as can the right hemisphere in the opposite situation (Figure 8.1).

These considerations lead to a review of the cerebral basis of attention, with which neglect is intimately connected. As reviewed in Chapter 2, attention can be divided into categories of selective, sustained, and directed. Selective and sustained inattention can be thought of as manifestations of bilateral neglect, and directed inattention is essentially synonymous with hemineglect. Although the acute confusional state is typically due to a widespread disturbance of the diffuse

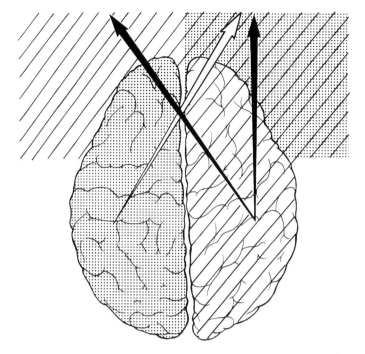

FIGURE **8.1.** *Schematic diagram of right hemisphere dominance for attention. Whereas the left hemisphere can only attend to the right side of extrapersonal space, the right hemisphere can attend to both sides.*

attentional system, it can also result from right parietal damage (Mesulam et al. 1976), and a frequent chronic sequel of such lesions is hemineglect. A network of frontal, parietal, and limbic structures in the right hemisphere mediating attention to the left hemispace has been postulated based on the observations of neglect patients with damage to all these regions (Mesulam 1981). These areas all appear to mediate specific aspects of directed attention: the frontal component, consisting of prefrontal cortex and the frontal eye field (Brodmann area 8), subserves motor exploration and visual scanning; the parietal component, including posterior and inferior parietal cortices, subserves sensory awareness; and the limbic component, centered in the cingulate gyrus, subserves motivational relevance (Mesulam 1981). Recent experimental and clinical observations have indicated that white matter tracts coursing through the right hemisphere also contribute to the neural network for directed attention (Bartolomeo, Thiebaut de Schotten, and Doricchi 2007), indicating that the connections between cortical areas are as important as the regions themselves.

SPATIAL DISORIENTATION

Individuals with lesions of the posterior right hemisphere often display a loss of familiarity with their environment. Obvious difficulties resulting from *spatial disorientation* may be evident when a patient is found to be unable to give directions to get home from the hospital or gets lost trying to find the nurses' station. Deficits detected on map completion tasks can alert the clinician to the possibility of a right hemisphere lesion, and an acute insult to the right parietooccipital region is often responsible (Fisher 1982). Cases such as these are particularly noteworthy in view of evidence from functional neuroimaging studies implicating the parietal lobes in the analysis of spatial properties of visual stimuli (Chapter 7); the right parietal component of this dorsal system may be essential in the maintenance of orientation for place. When combined with bifrontal pathology, spatial disorientation can lead to a delusional belief that an individual's locale has been duplicated and therefore exists simultaneously in two places: the syndrome of reduplicative paramnesia (Benson, Gardner, and Meadows 1976; Filley and Jarvis 1987). Similar pathology can also lead to another reduplicative condition called the Capgras syndrome, in which patients have the delusional belief that people have been replaced by impostors (Alexander, Stuss, and Benson 1979).

The loss of topographic familiarity has also been termed *environmental agnosia* (Landis et al. 1986). Loss of knowledge regarding the locations of familiar sites probably reflects either reduced access to cortical areas encoding this information (a disconnection between the perceived material and stored knowledge about it) or the actual obliteration of the encoded information. In the first case, it would be appropriate to consider the syndrome a specific agnosia, and in the second, the term amnesia would be more accurate. A specific environmental agnosia explaining loss of topographic familiarity is supported by evidence that affected patients can produce maps of their environment from memory even though they cannot recognize the setting when faced with the actual physical landscape (Landis et al. 1986).

DRESSING APRAXIA

Again, the use of the word apraxia is debatable in this context (Cummings and Mega 2003), but *dressing apraxia* is widely known as a specific impairment in dressing. When presented with the task of putting on a garment, especially with a sleeve turned inside out or other additional impediment, a patient so affected will have difficulty recognizing top and bottom, back and front, and so on. The deficit involves elements of both neglect and spatial disorientation (Brain 1941). Care should be taken not to mistake the motor dysfunction of hemiparesis, parkinsonism, or ataxia for dressing apraxia. In keeping with the importance of right parietal regions in spatial orientation, the location of lesions causing this impairment is usually the right parietal lobe (Cummings and Mega 2003). Difficulty with activities of daily living can clearly result from dressing apraxia.

APROSODY

Prosody refers to a paralinguistic element of language by which emotional or affective components are conferred on propositional language (Ross 1981, 2000; Ross and Monnot 2008). A simple sentence, such as "I went to the movies," may take on distinctly different meanings depending on whether the declaration is inflected by the speaker with elation, dejection, surprise, and the like. Prosody confers this affective coloration to the grammatical and semantic elements of language by the phenomena of melody, pauses, intonation, stresses, and accents. A confusing issue is that the term *linguistic prosody* is used to describe the intonations, stresses, and pauses of propositional language—features that are indicated in written language by commas, colons, semicolons, periods, and question marks—and this aspect of language enables the speaker to express, for example, a declarative versus an interrogative sentence. Thus our current topic is more precisely termed *affective prosody*, although in common usage this concept is referred to simply as prosody. Studies of patients with right hemisphere lesions have demonstrated that, whereas aphasia is absent, difficulties in producing (Ross and Mesulam 1979) and comprehending affective language (Heilman, Scholes, and Watson 1975) are commonly seen, and the syndrome of *aprosody* or aprosodia can then be diagnosed. In contrast to the left hemisphere's dominance for propositional language, therefore, these considerations lead to the notion that the right hemisphere is dominant for the affective components of language (Ross and Mesulam 1979; Ross 2000; Ross and Monnot 2008).

Related to prosody is *gesture* or body language, which provides another avenue for a speaker to communicate beyond the restraints of propositional language. In contrast to volitional movements intended to express symbols known as pantomime, impairment of which is linked to aphasia and left hemisphere damage, gesture is a right hemisphere function and its impairment is often associated with aprosody (Ross 2000). Taken together, prosody and gesture serve to embellish propositional language with emotional coloration that significantly expands the speaker's communicative power (Ross 2000). Figure 8.2 illustrates a right hemisphere network that is postulated to subserve prosody and gesture.

Detailed clinical studies of prosody and gesture have been pursued (Ross 1981, 2000; Ross and Monnot 2008). There appears to be a right hemisphere system for the mediation of prosody (Figure 8.2), which in general is a mirror image of the left hemisphere propositional language system reviewed in Chapter 5 (Figure 5.2). Focal lesions in this circuit often result in clinically detectable

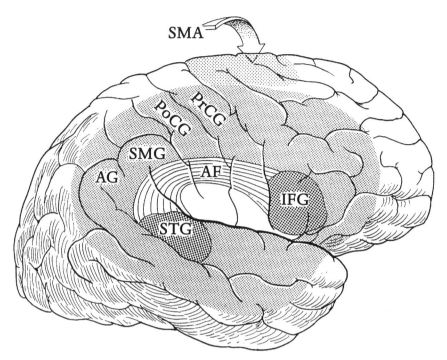

FIGURE **8.2.** *Lateral view of the right hemisphere showing areas important for prosody (IFG: inferior frontal gyrus; STG: superior temporal gyrus; AF: arcuate fasciculus; SMA: supplementary motor area; AG: angular gyrus; SMG: supramarginal gyrus; PrCG: precentral gyrus; PoCG: postcentral gyrus).*

aprosodic deficits that bear considerable similarity to the analogous syndrome that follows a similarly placed lesion of the left hemisphere (Table 8.2). Hence there are clinical examples of such entities as motor, sensory, conduction, global, and transcortical aprosodias that follow stroke or other right hemisphere lesions (Ross 1981, 2000; Ross and Monnot 2008) and that can be formally tested with an instrument called the Aprosodia Battery (Ross and Monnot 2008). Although the detection of the aprosody syndromes can be clinically challenging, aprosodic language is often a useful neurobehavioral sign of right hemisphere pathology. In theoretical terms, the study of aprosody has offered important insights regarding the means by which emotional dimensions enrich propositional language (Ross 2000; Ross and Monnot 2008).

Precise localization of the aprosodias has proven somewhat elusive. The prominence of right hemisphere lesions has been repeatedly confirmed, but beyond that, a strict mirror-image correspondence between the aphasias on

TABLE **8.2.** Aprosody syndromes

Aprosody Syndrome	Affective Aprosody			Gestures	
	Spontaneous	Repetition	Comprehension	Spontaneous	Comprehension
Motor	Poor	Poor	Good	Poor	Good
Sensory	Good	Poor	Poor	Good	Poor
Conduction	Good	Poor	Good	Good	Good
Global	Poor	Poor	Poor	Poor	Poor
Transcortical motor	Poor	Good	Good	Poor	Good
Transcortical sensory	Good	Good	Poor	Good	Poor
Agesic	Good	Good	Good	Good	Poor
Mixed transcortical	Poor	Good	Poor	Poor	Poor

Source: Ross 2000, 320.

the left and the aprosodias on the right has not been possible to document. However, lesions producing motor aprosody cluster in the areas of the right frontal operculum, and lesions leading to sensory aprosody are typically found in the right posterior temporoparietal region (Ross 2000). These aprosodies can be important clues to focal right hemisphere lesions and help guide further evaluation and treatment. Other aprosodias have a less secure localization, but the importance of the right hemisphere has held up well (Ross and Monnot 2008). The schema displayed in Figure 8.2 should be seen as a general guide to right hemisphere mediation of affective prosody, and it is clear that as further data appear, modifications in our understanding of the organization of this network can be anticipated.

AMUSIA

Music is a human capacity evident in all cultures, and its communicative power can exert profound effects on both performers and listeners. For neuroscientists, music also appeals as a fascinating phenomenon that must somehow be localized in the brain (Levitin 2006; Sacks 2007). However, the topic of musical representation in the brain is exceedingly complex, and inclusion of *amusia*, the acquired loss of musical skill, on the list of right hemisphere syndromes risks oversimplification. The many elements required for the successful performance or appreciation of music—pitch, duration, timbre, volume, tempo, melody, harmony, rhythm, and so on—are extremely diverse and do not allow convenient localization. Moreover, because musical mastery is relatively uncommon in the general population, few well-studied cases of amusia are available. Nevertheless, the traditional view that music is primarily a right hemisphere function (Milner

1962; Damasio and Damasio 1977) is accurate enough as a rough approximation to justify its discussion here.

One of the most striking demonstrations of right hemisphere participation in music is the unexpected production of intact singing in a musically competent individual with a left hemisphere lesion and nonfluent aphasia. Often as surprising to the patient as to the examiner, such an output clearly suggests the unimpaired musical capacity of the right hemisphere. Similarly, the aprosodic, amelodious speech characteristic of many patients with right hemisphere lesions is one of many suggestions that music may be associated with right hemisphere function. Studies using unilateral intracarotid amobarbital injection (the Wada test, or "reversible hemispherectomy") have supported these observations by finding that right carotid injection impairs singing ability much more than left (Gordon and Bogen 1974). This capacity of the right hemisphere may even be exploited by speech pathologists to help nonfluent aphasics intone speech in a melodious manner to improve their verbal output (Albert, Sparks, and Helm 1973; Norton et al. 2009).

Although the assumption of right hemisphere dominance for some aspects of music can be useful in some clinical settings, the representation of this highly complex activity is in fact bihemispheric, depending on the specific feature being considered (Gates and Bradshaw 1977). For example, rhythm, which is closely linked to language, is more a function of the left hemisphere, whereas melody is more associated with the right (Polk and Kertesz 1993). Both hemispheres undoubtedly participate, and functional neuroimaging studies have identified widespread neural networks dedicated to musical function (Sergent 1993). Studies performed in highly skilled musicians have suggested the interesting possibility of left hemisphere mediation of music in these individuals, while persons with no musical competence may process music more with the right hemisphere (Bever and Chiarello 1974). More recent data have revealed that perfect (absolute) pitch is associated with expansion of the planum temporale of the left superior temporal gyrus (Goldberger 2001).

A distinction between syndromes affecting musical execution—*expressive amusia*—and those affecting musical perception—*receptive amusia*—may be useful. It has been suggested that musical execution depends on the right hemisphere, whereas musical perception is a right hemisphere function in naïve listeners but a left hemisphere function in those who are musically sophisticated (Damasio and Damasio 1977). A shift of musical dominance may thus take place from the right to the left hemisphere as musical sophistication increases (Damasio and Damasio 1977).

Amusia is infrequent in clinical practice, but convincing cases have been reported after right temporoparietal infarct (McFarland and Fortin 1982), right frontotemporal degeneration (Confavreux et al. 1992), and right frontal lobe resection for relief of epilepsy (McChesney-Atkins et al. 2003). Whereas cases such as these clearly suggest a key role of the right hemisphere in music, caution is warranted since amusia has been seen in patients both with and without simultaneous aphasia and with either left or right hemisphere lesions (Brust 1980). The localization of musical function in all its variety thus remains uncertain, and more clarity in this area must await further research.

EMOTIONAL DISORDERS

The subject of *emotion* has not been pursued by behavioral neurologists as much as cognition, and until recently, psychiatry and psychology have been primarily charged with the task of addressing this imposing topic. Not long ago, in fact, under the influence of Freudian thinking in the early twentieth century, the study of the structure and function of the brain was thought to add little to the understanding of emotions. But this assumption has been seriously reconsidered, and much empirical evidence relating brain dysfunction with emotional disorders has appeared. In the past several decades, for example, new information has been presented on schizophrenia that increasingly implicates brain dysfunction, and the emerging field of neuropsychiatry has pursued productive investigation in this area. Similarly, taking advantage of the lesion method, many investigators have presented descriptions of emotional changes with structural brain disease, offering fascinating insights into the cerebral organization of emotional functions. Some of these syndromes will be discussed in the context of primary emotions mediated by the limbic system (Chapter 9), and in relation to frontal lobe supervision of emotion (Chapter 10), but we now consider the perplexing and often disabling disorders in emotional adjustment that can follow damage to right hemisphere structures (Price and Mesulam 1985; Cummings 1997).

As mentioned above, syndromes of the right hemisphere are difficult to characterize because of the nonverbal nature of the deficits. Yet patients with right hemisphere damage do manifest changes in emotion and personality that are as clinically significant as they are elusive to capture descriptively (Price and Mesulam 1985; Cummings 1997). These changes are evident in the realms of mood, affect, prosody, and interpersonal relations, and our understanding is still limited. Much of this complexity, however, can be put into context by a brief consideration of the neurology of emotion.

A firm consensus among neuroscientists holds that the limbic system (Chapter 9) plays a critical role in the primary emotions of the organism, ensuring such essential requirements of survival as feeding, aggression, flight, and reproduction. Instinctual drives result from the activity of the limbic system, which in humans has showed only modest enlargement in comparison with that of higher primates (Eccles 1989). The enormous expansion of the neocortex and its connections in humans, however, has enabled the operation of a much more elaborate behavioral repertoire (Eccles 1989), which is evident in the emotional as well as the cognitive realm. Here it is likely that the specifically human emotions are represented. Thus, humans share many features of emotional behavior with lower animals but, in addition, display unique behaviors that result from the interplay between the limbic system and the neocortical regions by which it is regulated.

The role of neocortical systems in the organization of emotion is still largely obscure. Clearly the neocortex mediates a variety of modulatory influences upon the limbic system that render human emotional behavior far more variable than the predictable stimulus-bound behavior of a lower animal in which the limbic system is anatomically more prominent. The frontal lobes, notably the orbitofrontal regions, provide a crucial element of neocortical control, and these areas can be viewed as essential to drive regulation (Chapter 10). In addition, the right hemisphere exerts a complementary although more subtle influence on limbic activity, enabling a richer and more variegated interpersonal life by means of such attributes as prosody, mood, affect, and socialization.

In Chapter 2, the Capgras syndrome (Alexander, Stuss, and Benson 1979), the Fregoli syndrome (Feinberg et al. 1999), and reduplicative paramnesia (Benson, Gardner, and Meadows 1976; Filley and Jarvis 1987) were introduced as delusional misidentification syndromes, and, not incidentally, theories of pathogenesis in all three implicate both the frontal lobes and the right hemisphere. Phenomenologically, these syndromes all share a disruption of the relationship between the self and other significant people, faces, and objects in the world. From these insights, it has been proposed that the right hemisphere, with a particular emphasis on the right frontal lobe, may be dominant for self-awareness (Feinberg and Keenan 2005). In this light, the newly introduced concept of *theory of mind* (Chapter 10) is also relevant: the ability to use one's beliefs, attitudes, and experiences to understand the mental states of others, or in essence, the capacity to be aware of another's mind (Stuss and Anderson 2004). These considerations readily recall the Freudian concept of ego, and whether the right frontal lobe can be considered as central to the organization of the ego must of course remain speculative (Feinberg and Keenan 2005). Still, the notion that the right hemisphere plays a special role in the representation of the self is an intriguing possibility that bears upon not only the cerebral basis of the emotions but also the nature of consciousness itself.

Although the important role of the right hemisphere in emotional life has ample documentation, the potential contributions of the left hemisphere have not been ignored and a lively debate has centered on the cerebral laterality of emotion. Much has been written over the years about the lateralization of higher functions in the brain, and substantial effort has been devoted to the explora-

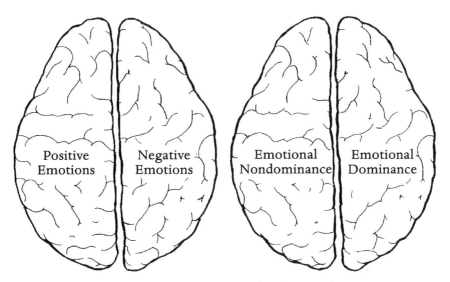

FIGURE **8.3.** *Two hypotheses concerning the hemispheric lateralization of emotion.*

tion of differences in the capacities of the left and right hemispheres (Springer and Deutsch 2001). Contrasts such as analytic-synthetic, logical-intuitive, and rational-emotional have been touted as defining traits, often with a paucity of supporting data. It is no surprise that a brisk controversy on the lateralization of emotion has developed. Two major theories on this contentious topic are current (Figure 8.3).

One idea, the *valence hypothesis*, is that the left hemisphere is primarily concerned with positive emotions, and the right with negative emotions (Sackheim et al. 1982). Thus happiness, for example, is mediated by the left side and sadness by the right. In contrast, the *right hemisphere hypothesis* holds that all higher emotion is mediated by the right hemisphere (Bear 1983). In this schema, right hemisphere dominance for emotion is postulated. Evidence from humans with left and right hemisphere lesions has been offered in support of both ideas, and the issue is unresolved. We shall review some key studies that are clinically helpful.

The valence hypothesis has been supported by several observations. For many decades it has been known that aphasic patients with left frontal lesions can exhibit a *catastrophic reaction* of profound depression and agitation (Goldstein 1948). This dramatic condition could plausibly reflect disruption of left hemisphere systems subserving positive emotions, so that the consequent clinical behavior expresses a strong depressive valence that arises from the intact right hemisphere where negative emotions are mediated. Thus the catastrophic reaction can be seen as a predictable response of the right hemisphere to the devastating loss of language fluency caused by the aphasia-producing lesion. More recently, data derived from stroke patients have refined this idea by identifying a specific link between left frontal infarcts and depression. Substantial evidence supports the association of left frontal cerebrovascular lesions and depression, even with lesions too far anterior to cause significant aphasia (Robinson et al. 1984; Narushima, Kosier, and Robinson 2003). These lesions may be either cortical or subcortical, and because affected patients can be effectively treated with tricyclic antidepressants (Starkstein and Robinson 1990), destruction of ascending catecholaminergic tracts may be responsible for this post-stroke depression. Other evidence supporting the valence hypothesis has been presented. In a report summarizing the consistent results of three studies, Sackheim and colleagues found that (1) pathologic laughter was associated with right hemisphere and crying with left hemisphere damage; (2) right hemispherectomy led to a euphoric mood change; and (3) epileptic foci causing gelastic (laughing) epilepsy were predominantly left-sided (Sackheim et al. 1982).

Much evidence favors the right hemisphere hypothesis. One early study showed that whereas patients with left hemisphere lesions displayed the catastrophic reaction of Goldstein, the typical response of right hemisphere–damaged patients was one of indifference, suggesting that the right hemisphere was more important for the mediation of emotional behavior (Gainotti 1972). Many patients with right cerebral lesions indeed show a remarkable passivity in the face of severe hemiplegia or other neurologic deficit. Also consistent with right hemisphere dominance for emotion are data from temporal lobe epilepsy patients: "emotive" personality traits such as emotionality, elation, and sadness correlate with right temporal seizure foci, while "ideative" traits such as a sense of personal destiny, philosophical interests, and humorlessness correlate with left temporal foci (Bear and Fedio 1977; Chapter 9). In another context, a large study of soldiers with unilateral penetrating head injuries found an association of affective disorders with right hemisphere and intellectual disorders with left hemisphere injuries (Lishman 1968). Other intriguing data come from many studies indicating a tendency for unilateral conversion disorder symptoms and signs to occur on the left side of the body (Galin, Diamond, and Braff 1977; Stern 1977; Ley 1980), implying that the psychic conflict that presumably leads to neurologic symptoms may be primarily organized by right hemisphere systems. Observations of stroke patients have highlighted an association of delusions with right hemisphere lesions, especially if superimposed on preexisting brain atrophy (Levine and Grek 1984), and mania has been observed after right anterior temporal lesions from traumatic brain injury (Starkstein et al. 1990). In normal young adults, Wittling and Roschmann (1993) found that films with both positive and negative emotional valence produced a stronger subjective response when presented to the right hemisphere than to the left. Finally, recent data from patients with frontotemporal dementia (Chapter 12) have informed this issue. In one study, socially undesirable behaviors such as aggression, alienation from family and friends, and sexual deviancy were more likely in patients with right than left hemispheric frontotemporal dementia (Mychack et al. 2001), and in another, behavioral disorders were more prevalent in frontotemporal dementia patients with right temporal lobe atrophy than in those with left temporal atrophy associated with semantic dementia (Chan et al. 2009). Thus, neurodegeneration of right frontal and temporal regions may interfere with normal social behavior to a greater extent than similar degeneration in the left hemisphere.

Other investigators have attempted to examine other aspects of interpersonal life requiring considerable emotional competence. One such domain is humor, and it has been found that patients with right hemisphere lesions demonstrate difficulty making an appropriate response to humorous material (Gardner et al. 1975). More recent data suggested a special role for the right frontal lobe in humor, as lesions of this region were most likely to disrupt the appreciation of humor and diminish associated responses such as laughter and smiling (Shammi and Stuss 1999). The ability to comprehend metaphor, the recognition of similarity between seemingly dissimilar terms, may also be impaired in right hemisphere–damaged patients, who tend to interpret metaphors literally as often as figuratively (Winner and Gardner 1977). Such patients have reduced metaphoric comprehension because of problems appreciating the connotative meaning of words—their allusions and implications—while retaining an understanding of their literal, denotative aspects (Brownell, Potter, and Michelow 1984; Brownell et al. 1990). These subtle deficits indicate that the right hemisphere plays an important role in the nuances of interpersonal relationships. Right hemisphere patients may fail to comprehend the subtle shades of meaning conveyed in conversation by humor, metaphor, and connotation.

Finally, some intriguing speculations have been based on data that dreams can be eliminated after damage to the right posterior hemisphere (Humphrey and Zangwill 1951). If dreams represent the "royal road to a knowledge of the unconscious activities of the mind" (Freud 1900, 647), then perhaps the right hemisphere may be the repository of the unconscious and the psychoanalytic concept of primary process thinking (Galin 1974). However, dreaming may not be solely a right hemisphere phenomenon. Others have suggested that whereas the visual aspect of dreams does implicate the right hemisphere, the narrative quality of dreams—which can persist in the absence of visual imagery—may implicate left hemisphere structures (Kerr and Foulkes 1981). Dreams may therefore result from the activation of widespread neurobehavioral networks in both hemispheres. This notion is consistent with evidence favoring bihemispheric participation in visual imagery (Chapter 7).

As a general rule, despite Freud's early interests in neurology and neuro-anatomy, the relevance of Freudian theory to behavioral neurology remains tenuous. Attractive as it may be to many in neuroscience (Kandel 1999), a secure linking of psychoanalytic theory to brain structure and function cannot be supported with existing knowledge. Nevertheless, the importance of the right hemisphere in many social and emotional operations—including those that may exist beneath conscious awareness—suggests that further elucidation of right hemisphere function will likely be assisted by the insights of psychiatrists, psychologists, and others who are experienced in the observation, analysis, and treatment of emotional disorders (Schore 1997). A collaboration between neurobiologists and mental health professionals will nowhere bear more fruit than in the study of the poorly understood human right hemisphere.

References

Albert, M. L., Sparks, R., and Helm, N. Melodic intonation therapy for aphasia. *Arch Neurol* 1973; 29: 130–131.

Alexander, M. P., Stuss, D. T., and Benson, D. F. Capgras syndrome: a reduplicative phenomenon. *Neurology* 1979; 29: 334–339.

Babinski, J. Contribution a l'étude des troubles mentaux dans l'hémiplégie organique cérébrale (anosognosie). *Rev Neurol* 1914; 27: 845–847.

Bartolomeo, P., Thiebaut de Schotten, M., and Doricchi, F. Left unilateral neglect as a disconnection syndrome. *Cereb Cortex* 2007; 17: 2479–2490.

Bear, D. M. Hemispheric specialization and the neurology of emotion. *Arch Neurol* 1983; 40: 195–202.

Bear, D. M., and Fedio, P. Quantitative analysis of interictal behavior in temporal lobe epilepsy. *Arch Neurol* 1977; 34: 454–467.

Benson, D. F., Gardner, H., and Meadows, J. C. Reduplicative paramnesia. *Neurology* 1976; 26: 147–151.

Bever, T. G., and Chiarello, R. J. Cerebral dominance in musicians and non-musicians. *Science* 1974; 185: 537–539.

Brain, W. R. Visual disorientation with special reference to lesions of the right cerebral hemisphere. *Brain* 1941; 64: 244–272.

Brownell, H. H., Potter, H. H., and Michelow, D. Sensitivity to denotation and connotation in brain-damaged patients: a double dissociation? *Brain Lang* 1984; 22: 253–265.

Brownell, H. H., Simpson, T. L., and Bihrle, A. M., et al. Appreciation of metaphoric alternative word meanings by left and right brain-damaged patients. *Neuropsychologia* 1990; 28: 375–383.

Brust, J.C.M. Music and language: musical alexia and agraphia. *Brain* 1980; 103: 367–392.

Chan, D., Anderson, V., Pijnenburg, Y, et al. The clinical profile of right temporal lobe atrophy. *Brain* 2009; 132: 1287–1298.

Confavreux, C., Croisile, B., Garassus, P., et al. Progressive amusia and aprosody. *Arch Neurol* 1992; 49: 971–976.

Critchley, M. *The Parietal Lobes*. London: Edward Arnold; 1953.

Cummings, J. L. Neuropsychiatric manifestations of right hemisphere lesions. *Brain Lang* 1997; 57: 22–37.

Cummings, J. L., and Mega, M. S. *Neuropsychiatry and Behavioral Neuroscience*. New York: Oxford University Press; 2003.

Damasio, A. R., and Damasio, H. Musical faculty and cerebral dominance. In: Critchley, M., and Henson, R. A., eds. *Music and the Brain*. London: William Heinemann; 1977: 141–155.

Denes, G., Semenza, C., Stoppa, E., and Lis, A. Unilateral spatial neglect and recovery from hemiplegia. *Brain* 1982; 105: 543–552.

Eccles, J. C. *Evolution of the Brain: Creation of the Self*. London: Routledge; 1989.

Feinberg, T. E., Eaton, L. A., Roane, D. M., and Giacino, J. T. Multiple Fregoli delusions after traumatic brain injury. *Cortex* 1999; 35: 373–387.

Feinberg, T. E., and Keenan, J. P. Where in the brain is the self? *Conscious Cogn* 2005; 14: 661–678.

Filley, C. M., and Jarvis, P. E. Delayed reduplicative paramnesia. *Neurology* 1987; 37: 701–703.

Fisher, C. M. Disorientation for place. *Arch Neurol* 1982; 39: 33–36.

Freud, S. *The Interpretation of Dreams*. Trans. and ed. Strachey, J. New York: Avon; 1900.

Gainotti, G. Emotional behavior and hemispheric side of lesion. *Cortex* 1972; 8: 41–55.

Galin, D. Implications for psychiatry of left and right cerebral specialization: a neurophysiological context for unconscious processes. *Arch Gen Psychiatry* 1974; 31: 572–583.

Galin, D., Diamond, R., and Braff, D. Lateralization of conversion symptoms: more frequent on the left. *Am J Psychiatry* 1977; 134: 578–580.

Gardner, H., Ling, P. K., Flamm, L., and Silverman, J. Comprehension and appreciation of humorous material following brain damage. *Brain* 1975; 98: 399–412.

Gates, A., and Bradshaw, J. L. The role of the cerebral hemispheres in music. *Brain Lang* 1977; 4: 403–431.

Geschwind, N. The apraxias: neural mechanisms of disorders of learned movement. *Am Sci* 1975; 63: 188–195.

Goldberger, Z. D. Music of the left hemisphere: exploring the neurobiology of absolute pitch. *Yale J Biol Med* 2001; 74: 323–327.

Goldstein, K. *Language and Language Disturbances*. New York: Grune and Stratton; 1948.

Gordon, H. W., and Bogen, J. E. Hemispheric lateralization of singing after intracarotid sodium amylobarbitone. *J Neurol Neurosurg Psychiatry* 1974; 37: 727–738.

Greene, J.D.W. Apraxia, agnosias, and higher visual function abnormalities. *J Neurol Neurosurg Psychiatry* 2005; 76 (Suppl V): v25–v34.

Hecaen, H., and Albert, M. L. *Human Neuropsychology*. New York: John Wiley and Sons; 1978.

Heilman, K. M., Scholes, R., and Watson, R. T. Auditory affective agnosia: disturbed comprehension of affective speech. *J Neurol Neurosurg Psychiatry* 1975; 38: 69–72.

Heilman, K. M., and Van Den Abell, T. Right hemisphere dominance for attention: the mechanism underlying hemispheric asymmetries of inattention (neglect). *Neurology* 1980; 30: 327–330.

Heilman, K. M., Watson, R. T., and Valenstein, E. Neglect and related disorders. In: Heilman, K. M., and Valenstein, E., eds. *Clinical Neuropsychology*. 4th ed. New York: Oxford University Press; 2003: 296–346.

Hier, D. B., Mondlock, J., and Caplan, L. R. Behavioral abnormalities after right hemisphere stroke. *Neurology* 1983; 33: 337–344.

Humphrey, M. E., and Zangwill, O. E. Cessation of dreaming after brain injury. *J Neurol Neurosurg Psychiatry* 1951; 14: 322–325.

Kandel, E. R. Biology and the future of psychoanalysis: a new intellectual framework for psychiatry revisited. *Am J Psychiatry* 1999; 156: 505–524.

Kaplan, E. Process and achievement revisited. In: Wapner, S., and Kaplan, B., eds. *Toward a Holistic Developmental Psychology*. Hillsdale, NJ: Lawrence Erlbaum; 1983: 143–156.

Kerr, N. H., and Foulkes, D. Right hemisphere mediation of dream visualization: a case study. *Cortex* 1981; 17: 603–610.

Kertesz, A., Nicholson, I., Cancilliere, A., et al. Motor impersistence: a right hemisphere syndrome. *Neurology* 1985; 35: 662–666.

Landis, T., Cummings, J. L., Benson, D. F., and Palmer, E. P. Loss of topographic familiarity: an environmental agnosia. *Arch Neurol* 1986; 43: 132–136.

Levine, D. N., and Grek, A. The anatomic basis of delusions after right cerebral infarction. *Neurology* 1984: 34: 577–582.

Levitin, D. J. *This Is Your Brain on Music: The Science of a Human Obsession*. New York: Dutton; 2006.

Ley, R. G. An archival examination of an asymmetry of hysterical conversion symptoms. *J Clin Neuropsychol* 1980: 2: 61–70.

Lishman, W. A. Brain damage in relation of psychiatric disability after head injury. *Br J Psychiatry* 1968; 114: 373–410.

McChesney-Atkins, S., Davies, K. G., Montouris, G. D., et al. Amusia after right frontal resection for epilepsy with singing seizures: case report and review of the literature. *Epilepsy Behav* 2003; 4: 343–347.

McFarland, H. R., and Fortin, D. Amusia due to right temporoparietal infarct. *Arch Neurol* 1982; 39: 725–727.

Mesulam, M.-M. A cortical network for directed attention and unilateral neglect. *Ann Neurol* 1981; 10: 309–325.

Mesulam, M.-M., Waxman, S. G., Geschwind, N., and Sabin, T. D. Acute confusional states with right middle cerebral artery infarctions. *J Neurol Neurosurg Psychiatry* 1976; 39: 84–89.

Milner, B. Laterality effects in audition. In: Mountcastle, V. B., ed. *Interhemispheric Relations and Cerebral Dominance*. Baltimore: Johns Hopkins University Press; 1962: 177–195.

Mychack, P., Kramer, J. H., Boone, K. B., and Miller, B. L. The influence of right frontotemporal dysfunction on social behavior in frontotemporal dementia. *Neurology* 2001; 56 (11 Suppl 4): S11–S15.

Narushima, K., Kosier, J. T., and Robinson, R. G. A reappraisal of poststroke depression, intra- and inter-hemispheric lesion location using meta-analysis. *J Neuropsychiatry Clin Neurosci* 2003; 15: 422–430.

Norton, A., Zipse, L., Marchina, S., and Schlaug, G. Melodic intonation therapy: shared insights on how it is done and how it might help. *Ann NY Acad Sci* 2009; 1169: 431–436.

Piercy, M., Hecaen, H., and de Ajuriaguerra, J. Constructional apraxia associated with unilateral cerebral lesions: left and right sided cases compared. *Brain* 1960; 83: 225–242.

Polk, M., and Kertesz, A. Music and language in degenerative disease of the brain. *Brain Cogn* 1993; 22: 98–117.

Price, B. H., and Mesulam, M. Psychiatric manifestations of right hemisphere infarctions. *J Nerv Ment Dis* 1985; 173: 610–614.

Robinson, R. G., Kubos, K. L., Starr, L. B., et al. Mood disorders in stroke patients: importance of location of lesion. *Brain* 1984: 107: 81–93.

Ross, E. D. Affective prosody and the aprosodias. In: Mesulam, M.-M., ed. *Principles of Behavioral and Cognitive Neurology*. 2nd ed. New York: Oxford University Press; 2000: 316–331.

———. The aprosodias: functional-anatomic organization of the affective components of language in the right hemisphere. *Arch Neurol* 1981; 38: 561–569.

Ross, E. D., and Mesulam, M.-M. Dominant language functions of the right hemisphere? Prosody and emotional gesturing. *Arch Neurol* 1979; 36: 144–148.

Ross, E. D., and Monnot, M. Neurology of affective prosody and its functional-anatomic organization in right hemisphere. *Brain Lang* 2008; 104: 51–74.

Sackheim, H. A., Greenberg, M. S., Weiman, A. L., et al. Hemispheric asymmetry in the expression of positive and negative emotions: neurologic evidence. *Arch Neurol* 1982; 39: 210–218.

Sacks, O. *Musicophilia: Tales of Music and the Brain*. New York: Alfred A. Knopf; 2007.

Schore, A. N. A century after Freud's project: is a rapprochement between psychoanalysis and neurobiology at hand? *J Am Psychoanal Assoc* 1997; 45: 807–840.

Sergent, J. Music, the brain and Ravel. *Trends Neurosci* 1993; 16: 168–172.

Shammi, P., and Stuss, D. T. Humour appreciation: a role of the right frontal lobe. *Brain* 1999; 122: 657–666.

Springer, S. P., and Deutsch, G. *Left Brain, Right Brain: Perspectives from Cognitive Science*. 5th ed. New York: W. H. Freeman; 2001.

Starkstein, S. E., Mayberg, H. S., Berthier, M. L., et al. Mania after brain injury: neuroradiological and metabolic findings. *Ann Neurol* 1990; 27: 652–659.

Starkstein, S. E., and Robinson, R. G. Depression following cerebrovascular lesions. *Semin Neurol* 1990; 10: 247–253.

Stern, D. Handedness and the lateral distribution of conversion reactions. *J Nerv Ment Dis* 1977; 164: 122–128.

Stuss, D. T., and Anderson, V. The frontal lobes and theory of mind: developmental concepts from adult focal lesion research. *Brain Cogn* 2004; 55: 69–83.

Weintraub, S., and Mesulam, M.-M. Right cerebral dominance in spatial attention. *Arch Neurol* 1987; 44: 621–625.

Winner, E., and Gardner, H. The comprehension of metaphor in brain-damaged patients. *Brain* 1977; 100: 717–729.

Wittling, W., and Roschmann, R. Emotion-related hemisphere asymmetry: subjective emotional responses to laterally presented films. *Cortex* 1993; 29: 431–448.

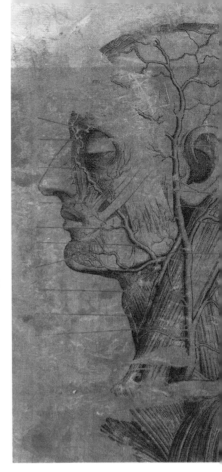

CHAPTER NINE

TEMPORAL LOBE SYNDROMES

It has been quipped that the Sylvian fissure, below which lies the temporal lobe and its various emotional functions, separates neurology from psychiatry. Although the position of this book is that any distinction between neurology and psychiatry will ultimately prove artificial (Chapter 1), it is true that a wide spectrum of emotional states, normal and otherwise, have been specifically linked with the temporal lobe and the limbic system to which it is intimately related. These regions therefore present a unique opportunity for exploration of emotional phenomena mediated by the brain. The cognitive domains of memory (Chapter 4) and language (Chapter 5) are closely linked with the temporal lobe and can be considered together with this chapter, which will concentrate on *temporal lobe epilepsy* (TLE). This illness is not only a common type of seizure disorder with a host of neurologic and psychiatric implications, but it also opens a particularly revealing window into brain-behavior relationships. TLE has long been postulated to produce a plethora of emotional changes that remain controversial today, and considering this disorder illuminates the challenge as well as the potential utility of examining emotional functions in the context of disturbed brain activity.

This book is largely concerned with cognition, the operations of which are simpler to define and measure than those of emotion. The neuroanatomy of cognition is similarly less complex than that of emotion. The previous chapter considered emotion in the context of right hemisphere disorders, and theories regarding the cerebral lateralization of emotion were reviewed. The chapter to follow will examine emotion (and cognition) in the setting of frontal lobe function. There remain, however, what have been called the "primary" emotions: pleasure, pain, desire, anger, ecstasy, fear, and the like (Ross, Homan, and Buck 1994; Dalgleish 2004). These powerful, basic feelings now direct our attention to the temporal lobes and their limbic connections.

THE LIMBIC SYSTEM

The term limbic derives from the Latin word for "border" (*limbus*), and it was Paul Broca who first introduced the idea of a limbic lobe (Broca 1878). Today the *limbic system* is commonly regarded as a network of interconnected structures in the temporal lobes and diencephalon strongly associated with aspects of emotion and memory. The components and boundaries of the limbic system are still not universally agreed upon (Brodal 1981; Nolte 2002), but the concept has served well to organize neurobiological thinking. The limbic system is in fact a key evolutionary structure, as it is similar in all vertebrates (Nolte 2002) and subserves basic emotions across species and cultures (Dalgleish 2004). The similar emotional expressions of humans and other animals famously noted by Charles Darwin (1872) are primarily related to the limbic system and its function, although humans have a far more elaborate emotional repertoire because of additional cerebral structures enabling more nuanced behavior.

Broca (1878) pointed out that a ring of cortical gyri around the upper brainstem—the cingulate gyrus, parahippocampal gyrus, and hippocampus—was characterized by phylogenetically primitive cortex. In other vertebrates, these areas are more associated with olfaction, but as this sensory modality is less important in humans, limbic function in humans is widely acknowledged to mean emotion and memory. Although lesions confined to the limbic system as a functional unit are rare, a well-studied case of bilateral limbic destruction from herpes simplex encephalitis does indeed demonstrate selective deficits in these domains; the patient had profound change in personality and affect and also dense amnesia (Feinstein et al. 2010).

In 1937, James Papez published an influential paper on the putative anatomic basis of emotion. Papez (1937) proposed that an interconnected circuit of limbic structures—the hippocampus, fornix, mammillary body, mammillothalamic tract, anterior thalamic nucleus, cingulate gyrus, and parahippocampal gyrus—mediates both the experience and expression of emotion (Figure 9.1).

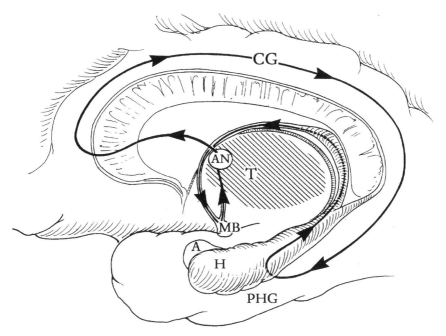

FIGURE **9.1.** *The Papez circuit (H: hippocampus; A: amygdala; PHG: parahippocampal gyrus; CG: cingulate gyrus; AN: anterior nucleus of thalamus; MB: mammillary body; T: thalamus).*

This *Papez circuit* has come to be regarded as playing a critical role in the primary emotions that form the basis of instincts or drives—feeding, aggression, flight, and sexual reproduction. The experience of these emotions was postulated as occurring in the cingulate gyrus, and their expression through the hypothalamus. More than seventy years of study have generally supported Papez's ideas, and the concept of the Papez circuit has anchored research efforts concerned with the cerebral basis of both emotion and memory.

In 1949, Paul MacLean extended Papez's formulation to include additional brain areas important to the phenomena of emotion: the amygdala, the septum and adjacent basal forebrain, the nucleus accumbens in the striatum, and the orbitofrontal cortex (MacLean 1949). MacLean thus gave further support to the assignment of emotional functions to the limbic system and stimulated much productive investigation by emphasizing the importance of additional regions. For our purposes, five major structures and their connections will serve as the most critical limbic components. These structures are the hippocampus and septal region, which have major roles in memory (Chapter 4), and the amygdala, cingulate gyrus, and hypothalamus, which are prominently involved in emotion. Of particular importance with regard to this chapter are the hippocampus and amygdala, structures that serve as dedicated limbic centers for memory and emotion. These regions are intimately linked, both neuroanatomically and functionally.

The amygdala is a limbic nucleus adjacent to the hippocampus that has recently been envisioned as an evaluator of stimuli entering the brain for assessment of their emotional significance (LeDoux 1987, 1994, 2007). Receiving both interoceptive (e.g., hunger, thirst) and exteroceptive (e.g., vision, audition, somatic sensation) information, the amygdala functions to assign appropriate emotional significance to these stimuli. Thus the amygdala is immediately interposed between the arrival of information and its further cerebral processing. This evaluation by the amygdala may occur in a nonconscious fashion, so that "emotional memories" may be the conscious products of unconscious activity (LeDoux 1994). The implications of such unconscious learning are substantial, as some of the most vexing problems in psychiatry involve a learned pattern of fear that occurs without conscious awareness and can be very difficult to reverse (LeDoux 1994). As research on the amygdala has continued to evolve, fear has been shown to be the emotion most firmly associated with amygdala function, and growing evidence supports the view that dysfunction of the amygdala is implicated in psychiatric disorders in which the experience of fear is central—anxiety, panic disorder, and post-traumatic stress disorder (LeDoux 2007)

The close proximity of the hippocampus to the amygdala is not a neuroanatomic coincidence; the same information that reaches the amygdala is also presented to the hippocampus, so that the possibility of storing the material for future use is apparent. The hippocampus and amygdala operate as a unit that synergistically forms memories of emotionally significant events (Richter-Levin and Akirav 2000). Thus the commonplace observation that memories of highly charged emotional experiences will be more easily formed and less readily dislodged than memories with little emotional significance is not surprising. The convergence of major sensory streams, most importantly from the visual cortices, into the limbic system facilitates the rapid processing of incoming information and enables a prompt and appropriate response by the individual.

Let us consider the implications of this arrangement. Successful existence depends on adaptive mechanisms whereby the organism can react appropriately to environmental contingencies. Thus evaluation of the emotional relevance of sensory information is clearly indispensable. In the case of a visual stimulus, for example, the information may be neutral (e.g., a stranger passing by on the street), positive (e.g., an appealing view of nature), or negative (e.g., a rapidly approaching truck). The amygdala, in the assessment of these possibilities, can operate in a variety of ways. First, there may be no further processing at all, as a neutral stimulus requires no action. Second, there may be the admittance of the stimulus into awareness, a process that contributes to the experience of emotion. While the cingulate gyrus likely participates in the experience of emotion, as Papez thought, the amygdala also clearly plays a key role (Gloor et al. 1982; Richter-Levin and Akirav 2000; Dalgleish 2004; Le Doux 2007). Third is the option of emotional expression, which may be displayed by an immediate reaction to the information received, such as enjoying an idyllic scene or rapidly evading the approach of the truck. Behavior such as this involves limbic connections to motor systems (Damasio and Van Hoesen 1983) and the autonomic (Bard 1928) and endocrine (Herman et al. 2005) effectors mediated within the hypothalamus (Dalgleish 2004). In this regard, evidence also supports an immunoregulatory function of the hypothalamus (Brooks et al. 1982), which may help explain a variety of psychosomatic diseases (MacLean 1949). Finally, the information evaluated by the amygdala may be transferred by its connections to the hippocampus for storage in memory; thus, for example, one might learn that a particular site has a pleasant view of nature or is potentially dangerous because of fast-moving vehicles. Just as the hippocampus is crucial for enabling the storage of information in the brain, the amygdala is the primary region for assessing whether information when first encountered requires such storage (Richter-Levin and Akirav 2000). In the case of horrific experiences such as those associated with prolonged abuse, torture, or combat, it may be that fear becomes tightly stored in the amygdalae, such that the memories may prove clinically problematic for years to come as post-traumatic stress disorder, anxiety, or panic (LeDoux 2007).

Although its neuroanatomy is difficult, the limbic system can thus usefully be considered as a functional unit. There is clinical and theoretical utility in viewing this network as critical to primary emotions and therefore to entities such as instincts, drives, pain, pleasure, fear, and anger. A complete portrait of the human neurobehavioral repertoire must include the limbic system as it performs its fundamental emotional functions and interacts with cerebral areas devoted to the subtleties of emotion and the various domains of cognition.

TEMPORAL LOBE EPILEPSY

The temporal lobes have intimate neuroanatomic relationships with the limbic system and, as we have seen, are closely involved with emotional behavior. The inseparable topography of the temporal lobes and deeper limbic structures helps explain why patients with temporal lobe lesions often manifest disorders that come to the attention of psychiatrists as readily as to neurologists. As discussed above, a key neuroanatomic feature is that all major sensory systems—vision, audition, and somatic sensation—are projected to the anterior temporal lobes and thence to the amygdala and hippocampus. This funneling of sensory input into emotionally relevant parts of the brain has implications for TLE, the major disease that has been examined in this regard. Not only is this a very common type of seizure disorder and thus accessible to detailed study, but it represents a specific cerebral lesion whose neurobehavioral manifestations can be readily examined (Waxman and Geschwind 1975).

The topic of epilepsy introduces a different conceptual approach to brain-behavior relationships than considered heretofore. The majority of syndromes in behavioral neurology begin with a focal lesion or degenerative process in a given cerebral region or network of structures that then permits the observation and analysis of deficits that may occur. In contrast to the "negative" signs and symptoms produced by a stroke, such as hemiparesis, hemianopia, aphasia, and neglect, epilepsy results in "positive" features, such as convulsive muscle activity or a visual hallucination, as a result of its characteristic irritative nature. Thus an entirely novel set of clinical phenomena can inform thinking about the representation of behavior in the brain. Deficits can now be related to the abnormal excitability of brain tissue as well as to its destruction.

A *seizure* is a clinical event caused by a paroxysmal and excessive discharge of cortical neurons. The term *epilepsy*, or alternatively *seizure disorder*, refers to a condition of recurrent, unprovoked seizures. In contrast, provoked seizures, such as those caused by alcohol withdrawal, do not justify the term seizure disorder because seizures in affected individuals would not occur in the absence of alcohol intoxication. Standard classifications of epileptic seizures, the most useful of which came from the International League Against Epilepsy in 1989 (Commission of ILAE 1989), generally divide seizure disorders into two broad categories: *generalized*, meaning that the entire brain is affected in a widespread fashion, and *partial*, implying the existence of a localized area of irritability from which the seizure originates. Partial seizures are further divided into *simple partial seizures* and *complex partial seizures*, the difference being that consciousness is preserved in the former but impaired or lost in the latter (Commission of ILAE 1989). Complex partial seizures most often result from an irritable focus in one of the temporal lobes. However, since epileptic foci causing complex partial seizures can be found outside the temporal lobes—in the frontal lobes, for example—the term TLE is not synonymous with complex partial epilepsy. Similarly, the older term psychomotor epilepsy cannot be equated with complex partial seizures and has in fact fallen from common use.

From the viewpoint of behavioral neurology, the problem of TLE serves to focus attention on the neuroanatomic interdigitations of the temporal lobes and the limbic system, a feature that has major neurobehavioral implications (Schomer et al. 2000). Some authorities in fact prefer the term temporolimbic epilepsy to TLE, but the familiarity of the latter term commends its use in this book. In addition to its theoretical interest, TLE is an important neurologic problem in its own right; it is the most common type of partial seizure, and perhaps 1 million people are afflicted with TLE in the United States alone (Schomer et al. 2000). That the illness is likely underreported only adds to the clinical importance of this difficult but neurologically instructive disorder (Schomer et al. 2000).

The phenomenology of TLE will be useful to consider. An important point to keep in mind is that TLE patients can manifest behavioral changes of three general types: ictal, postictal, and interictal. These categories respectively describe the period during a seizure, immediately following a seizure, and between seizures. *Ictal* phenomena include a variety of alterations in the motor, sensory, autonomic, cognitive, and emotional realms. A complete listing of these clinical features is beyond the scope of this book, but some of the more notable include focal jerking of the limbs and complex automatisms, focal numbness or discomfort, and an unpleasant "rising" sensation in the abdomen. Cognitive experiences include illusions such as micropsia, macropsia, and metamorphopsia and hallucinations in any sensory modality, frequently olfactory or gustatory (Currie et al. 1971). In addition, there are often cognitive alterations such as the "dreamy state" first described by Hughlings Jackson more than a century ago (1888–1889), which may be accompanied by illusions of familiarity (déjà vu) or unfamiliarity (jamais vu; Bancaud et al. 1994). Finally, emotional experiences are well described, the most common of which are fear and depression (Williams 1956). From knowledge now available (LeDoux 2007), the prominence of fear in the phenomenology of these seizures implies that seizure activity extends to the amygdala. Rare cases of ictal euphoria have also been reported, and it has been suggested that Fyodor Dostoyevsky may have experienced this emotion as ecstatic seizures and described the experience through his Prince Myshkin in *The Idiot* (Alajouanine 1963). *Postictal* alterations are typically those of a lethargic acute confusional state (Chapter 3), which are typically transient, and recently the syndrome of *postictal psychosis* has attracted more attention (Kanner 2000). *Interictal* changes, however, are those that can be observed during the great majority of the patient's life when seizures are not occurring. These proposed changes in personality and behavior are controversial but intriguing.

Before considering the specific neurobehavioral implications of TLE, a brief review of the psychiatric complications of epilepsy in general is in order. The history of epilepsy has unfortunately been blemished by uninformed prejudice, so much so, for example, that demonic possession was once believed to be its cause. Even today, the word *epilepsy* may connote the presence of negative traits in affected persons that many other diseases do not elicit. It is in this setting that attributions of behavioral changes in patients with TLE must be viewed. There is an understandable tendency for many clinicians to minimize the psychiatric or behavioral alterations that may be present, as it is thought that epileptics already have problems enough without the additional burdens of supposed undesirable character traits. Whereas stigmatization of any kind is clearly odious, it is nevertheless true that people with epilepsy, and particularly TLE, seem to have a special predilection for psychiatric morbidity.

A broad consensus exists that psychiatric disorders are more prevalent in people with epilepsy than in the population as a whole (Garciá-Morales, de la Peña Mayor, and Kanner 2008). Patients with TLE are especially vulnerable to psychiatric illness, as they are at higher risk for depression, anxiety, and psychosis than those with other types of epilepsy (Garciá-Morales, de la Peña Mayor, and Kanner 2008). The reasons for the increased prevalence of these disorders in TLE are doubtless manifold, including psychosocial factors and medication effects as well as the impact of the specific brain pathology, epileptogenesis, and localization (Devinsky and Najjar 1999). It has also been pointed out that, as the population ages and life expectancy is increased by improved public health measures and medical care, the impact of aging on the epileptic brain may introduce new and unexpected factors that could negatively impact outcome (Hermann, Seidenberg, and Jones 2008). The most important clinical point to emphasize is that psychiatric illness—however its cause is explained—should not be overlooked in the care of individuals with clinical dysfunction arising from a cortical lesion, the seizures it generates, and a host of associated problems that may compromise the quality of life.

For behavioral neurologists, however, the issue of personality and behavioral changes that may occur as a consequence of chronic temporolimbic irritability remains important from a theoretical perspective. Unlike the destructive lesions that have provided the basis for brain-behavior correlations in behavioral neurology, irritative lesions offer a different avenue of study based on the chronic effects of an abnormal discharge in identifiable cerebral areas. The potential sequelae of a lasting increase in cortical irritability deserve attention by virtue of the potential insights they may disclose about the brain and the organization of emotion. In fact, it has been contended that study of the behavioral features of TLE—founded on the characterization of clinical phenomena associated with a known irritative cerebral lesion—is the most promising area of psychiatry (Geschwind 1977). Thus it might make more sense to inquire how schizophrenia resembles interictal changes in TLE rather than the opposite (Geschwind 1977). Given the highly productive assumption that all behavior is mediated by the brain, these ideas are as intriguing now as when first proposed.

In the sections below, two behavioral syndromes that have attracted attention as being in some way related to the irritable cortical lesion of TLE will be considered. These syndromes have been and remain contentious and are presented to illustrate how thinking about TLE has evolved from what is known of the temporal lobes and the limbic system lobes in the organization of human behavior.

Psychosis in Temporal Lobe Epilepsy

Psychosis can occur in patients with many types of seizure disorder but most often in those with TLE (Kanner 2000; Nadkarni, Arnedo, and Devinsky 2007). Psychosis can appear in many clinical settings where TLE is implicated; most crucial for our purposes is the development of psychosis as an ictal phenomenon, in the postictal period, and as an interictal syndrome that may take years to develop. We shall consider these various psychoses in turn.

Ictal psychosis refers to a psychotic episode caused by nonconvulsive status epilepticus (Kanner 2000). The seizures are often complex partial seizures of temporal lobe origin, but simple partial and absence status epilepticus may also produce psychosis during the period of the epileptic discharges. Motor and other more familiar signs of status epilepticus may be lacking, but electroencephalography is diagnostic in these cases, and pharmacological control of the seizures promptly resolves the psychosis.

Postictal psychosis has attracted attention in recent years as it has been recognized with increasing frequency (Kanner 2000). TLE is the most frequent seizure type associated with this syndrome, and patients can experience the onset of either delusional or manic psychosis beginning within three days of a complex partial seizure or cluster; the period of psychosis lasts an average of fifteen hours (Kanner 2000). Although the mechanism of postictal psychosis remains uncertain, the presence of bilateral independent epileptic foci is known to enhance the risk of this disorder (Kanner 2000). These observations imply that the substantial modification of temporolimbic circuitry implied by lasting, bilateral irritable lesions of the limbic cortex may disturb cognition to the degree that the additional stress of seizures can result in the unmasking of an overt thought disorder. Prevention of seizures is an obvious treatment strategy but may be difficult in view of the relative intractability of these seizures. Thus patients may experience numerous psychiatric encounters or even hospitalizations related to postical psychosis before effective management to prevent seizures can be instituted. As for managing the psychosis itself, neuroleptic drugs with low potential to lower the seizure threshold can be considered (Kanner 2000).

Finally, we consider the interictal psychosis of TLE, or what has been called the *schizophreniform psychosis* of epilepsy or schizophrenia-like psychosis (Slater and Beard 1963). The possibility that chronic psychosis of this type might result from the psychological effects of a chronic and unpredictable illness should be borne in mind, but several studies have supported an association between psychosis and TLE but not other seizure types (Perez and Trimble 1980; Ramani and Gumnit 1982; Mendez et al. 1993). In the original account of this syndrome, individuals with TLE were found to develop an interictal delusional and paranoid psychosis with auditory hallucinations some fourteen years after their seizures began (Slater and Beard 1963). The patients were thought not to have schizophrenia because there was a preservation of affect, for which the modifier "schizophreniform" was therefore chosen, and because a significant family history of schizophrenia and premorbid schizoid traits in the TLE patients was absent. Further study has suggested that this psychosis may eventually occur in 7 to 15 percent of TLE patients (Trimble 1983; Schomer et al. 2000). Of note, the outcome of this disorder is thought to be more favorable than in schizophrenia itself (Kanner 2000; Schomer et al. 2000) The mechanism has been hypothesized to be a *sensory-limbic misconnection* induced by an abnormal area of irritable cortex in the temporal lobe (Schomer et al. 2000). Hence a neutral sensory experience becomes associated with an incongruent feeling state; for example, an innocent stranger seen incidentally might be interpreted as threatening to a TLE patient, and the result would be clinically significant paranoia. Seizure control does not, unfortunately, improve this problem, implying that chronic subictal irritability is sufficient to maintain the syndrome. In fact, seizure control may actually exacerbate the psychosis, a phenomenon described half a century ago as *forced normalization* (Landolt 1953) or *alternative psychosis* (Kanner 2000). This curious syndrome implies an inverse relationship between seizures and psychosis, as though "forcing" the electroencephalogram to be normal by control of seizures facilitates the emergence of psychosis.

These considerations lead to a discussion of schizophrenia itself, the disabling disorder of unknown etiology that typically begins in young adulthood and results in chronic mental dysfunction (Carlsson 1988; Rapoport et al. 2005; van Os and Kapur 2009). Although not traditionally considered a neurologic disease, schizophrenia has been studied as a brain disease since the time of Alzheimer and Kraepelin, and evidence continues to mount supporting the presence of brain dysfunction. The response of schizophrenic patients to neuroleptic drugs that act by blockade of dopamine receptors (Carlsson 1988; van Os and Kapur 2009) has suggested for many years that the illness involves more than psychodynamic factors, and the emphasis on psychopharmacology has clearly improved the treatment of schizophrenia for countless individuals who no longer need face the prospect of long-term psychiatric hospitalization. More recently, it has become clear that the majority of patients with schizophrenia have cognitive impairment, most notably in episodic memory and speed of processing, that may actually be more important in determining functional disability than psychotic features (Palmer, Dawes, and Heaton 2009). These lines of inquiry have clearly established that schizophrenia is a neurobiological disease with a major impact on cognition. However, the etiology and pathogenesis of the disease remain obscure, and the investigation of potential structural abnormalities in the brain has therefore attracted much interest. An early clue to the possibility of brain pathology was the insight that many neurological illnesses, including TLE, may be associated with psychosis resembling schizophrenia (Davison and

Bagley 1969). If neurologic disease can cause psychosis, it was reasoned, then schizophrenia itself may have a neurologic basis. This hypothesis, compelling as it is, has been difficult to substantiate in the past because of the notoriously elusive neuropathology of schizophrenia.

The debate over the details, and even the existence, of neuropathological features of schizophrenia has not been completely settled even after a century of investigation, particularly at the microscopic level (Harrison 1999). It does appear, however, that brain weight is on average decreased in schizophrenic subjects (Roberts 1991), an observation consistent with numerous imaging studies using computed tomography (CT) and magnetic resonance imaging (MRI) that document brain atrophy in many schizophrenics as reflected in lateral and third ventricular enlargement (Hyde and Weinberger 1990). Current thinking supports the proposal that the primary pathology lies in the frontal and temporal lobes (Iritani 2007) and that both gray matter (Harrison 1999; Iritani 2007) and white matter are implicated (Kubicki, McCarley, and Shenton 2005; Walterfang et al. 2006). At the microscopic level, subtle alterations in neuronal cell bodies, synapses, and glial cells can all be seen, consistent with a disorder of both synapses and connectivity (Harrison and Weinberger 2005). Although the etiology of the disease is unknown, genetic factors are now being strongly considered (Harrison and Weinberger 2005; Walterfang et al. 2006).

Considerable evidence implicates the temporal lobes. Volumetric studies of postmortem schizophrenic brains have found reductions in the hippocampus, amygdala, and parahippocampal gyrus in comparison with controls (Bogerts, Meertz, and Schonfeldt-Bausch 1985). MRI studies have also noted selective temporolimbic involvement in schizophrenia; for example, Suddath and colleagues (1989) found reductions in hippocampal and amygdala volume. Furthermore, volume loss in the left medial temporal region has been found specifically in schizophrenics (Shenton et al. 1992), and positron emission tomography (PET) studies have found increased metabolism in this same region (Friston et al. 1992). The findings implicating left temporal dysfunction in schizophrenia are particularly intriguing in light of the suggestion of Flor-Henry (1969) that this region may be important in the psychosis of TLE. There are also data supporting frontal lobe dysfunction, although of a different sort. Functional neuroimaging studies have consistently documented bifrontal decreases in metabolic activity, a pattern called "hypofrontality" in schizophrenics, especially when they are given tasks that emphasize prefrontal function (Weinberger and Berman 1988). Considerable evidence supports the idea of reduced metabolic activity in the prefrontal cortex (Carpenter and Buchanan 1994). Most recently, interest has turned to the temporal and frontal lobe white matter in schizophrenia, where abnormalities in the organization of myelinated tracts and in oligodendrocytes have been identified by neuropathological and neuroimaging studies (Kubicki,

McCarley, and Shenton 2005; Walterfang et al. 2006). The idea of cerebral disconnection may thus be relevant to the pathogenesis of schizophrenia.

In light of this evidence, a combination of temporal and frontal dysfunction has been proposed to explain the essential abnormalities of schizophrenic behavior (Weinberger 1986; Gold and Weinberger 1995). Pertinent to this argument is a distinction between the "positive" clinical features of schizophrenia—delusions, hallucinations, and thought disorder—and the "negative" or "deficit" features—apathy, amotivation, and social withdrawal (Crow 1980). Positive characteristics are seen as related to limbic dysfunction and the likelihood of mesolimbic dopaminergic overactivity, while negative or deficit manifestations are viewed as due to prefrontal dysfunction, possibly associated with a deficiency of dopamine (Weinberger 1986). Neuropsychological evidence now confirms that cognitive dysfunction is more severe in the latter form (Palmer, Dawes, and Heaton 2009). Both neurobehavioral profiles may be sequelae of a neurodevelopmental lesion, the effects of which become manifest when maturity is reached (Weinberger 1986; Rapoport et al. 2005). Recent findings suggest that several susceptibility genes play a role in the pathogenesis of schizophrenia (Harrison and Weinberger 2005).

The relationship of the TLE schizophreniform psychosis to schizophrenia itself becomes clearer in this light. In both conditions there is a putative dysfunction of temporolimbic systems—albeit by different processes—that leads to psychiatric phenomena. If this theory proves to be correct, then perhaps the observed clinical phenomena are not psychiatric at all but rather neurologic (Chapter 1). It is intriguing to speculate that the preservation of affect that characterizes schizophreniform psychosis may be due to the absence of frontal pathology in TLE, a situation that unfortunately does not obtain in many schizophrenics. Neuropsychological studies comparing schizophrenics and TLE patients with left- and right-side foci have disclosed that schizophrenics have attentional deficits suggesting frontal dysfunction that are not seen in either TLE group (Gold et al. 1994). Indeed, the negative features of schizophrenia bear a close similarity to the clinical deficits of dorsolateral frontal lobe dysfunction (Chapter 10).

TEMPORAL LOBE EPILEPSY PERSONALITY

Another important area of study in TLE—one that has engendered a great deal of debate—is the idea of a TLE personality. This concept may have generated as much dispute as any in neurology for several generations. To begin, it is a common experience in the care of persons with TLE, for example, that patient contacts can be prolonged, intense, and affectively charged, a feature referred to as *viscosity*. The personality alterations many have noted in TLE have been considered to be an interictal disturbance that, like the schizophreniform psychosis, can follow the onset of seizures, and a wide variety of personality traits have been suggested to be potential sequelae of the chronic effects of a seizure disorder secondary to a temporal lobe epileptic focus. For a century or more, case reports have described these changes in anecdotal and subjective fashion, and indeed the imprecision of the personality features remains one of the major criticisms of the theory. Moreover, as with schizophreniform psychosis, the influence of psychologic factors cannot be ignored. However, in the past several decades more systematic study has been undertaken, and a growing research literature has illuminated this difficult issue to a considerable extent.

A major figure in this controversy was Norman Geschwind, who in 1974 presented his ideas on personality change in TLE in a notable lecture at Harvard Medical School (Geschwind 1977). In 1975, Geschwind collaborated with Waxman and formally proposed the existence of an interictal behavioral syndrome in TLE patients characterized primarily by changes in sexual behavior, religiosity, and hypergraphia (Waxman and Geschwind 1975). A deepening of emotions manifesting in an intensification of interpersonal contact was considered central to this syndrome. Although such a formulation may be seen as calling undue attention to questionable personality changes in persons afflicted with a chronic disease, it is important to point out that Waxman and Geschwind commented that these traits were not necessarily maladaptive and that their observations identified a behavioral *change* rather than a behavioral *disorder* (Waxman and Geschwind 1975).

Pursuing this line of inquiry, Bear and Fedio (1977) undertook a literature review and determined that eighteen different personality traits (Table 9.1) could be linked with TLE; they then studied a large group of TLE patients in comparison with normal and neurological control subjects and found that all eighteen traits were more prevalent in the TLE group (Bear and Fedio 1977). These traits were evident both on patients' self-report and on the reports of observers. Although criticized methodologically, the study of Bear and Fedio (1977) has endured as a pioneering attempt to describe the personality features of TLE patients. In clinical practice, detection of all these personality alterations in the same patient is unusual, but finding a cluster of them in one person is not; as in other medical syndromes, individual variability in clinical expression is the rule rather than the exception. To sum up, Geschwind and his colleagues ignited an enduring debate about the topic that continues to this day, and to recognize his importance in sparking this discussion, the appellation of the syndrome as the *Geschwind syndrome* has been advanced (Benson 1994).

The mechanism of the syndrome has not been firmly elucidated but has been postulated to be a *sensory-limbic hyperconnection*, whereby sensory experience is suffused with excessive emotional coloration because of the increased electrical activity in the temporal lobe that enhances the neuroanatomic and

TABLE **9.1.** Personality traits associated with temporal lobe epilepsy

Inventory Trait	Reported Clinical Observations
Emotionality	Deepening of all emotions, sustained intense affect.
Elation, euphoria	Grandiosity, exhilarated mood; diagnosis of manic-depressive disease.
Sadness	Discouragement, tearfulness, self-deprecation; diagnosis of depression, suicide attempts.
Anger	Increased temper, irritability.
Aggression	Overt hostility, rage attacks, violent crimes, murder.
Altered sexual interest	Loss of libido, hyposexuality; fetishism, transvestism, exhibitionism, hypersexual episodes.
Guilt	Tendency to self-scrutiny and self-recrimination.
Hypermoralism	Attention to rules with inability to distinguish significant from minor infractions; desire to punish offenders.
Obsessionalism	Ritualism; orderliness; compulsive attention to detail.
Circumstantiality	Loquacious, pedantic; overly detailed, peripheral.
Viscosity	Stickiness; tendency to repetition.
Sense of personal destiny	Events given highly charged personalized significance; divine guidance ascribed to many features of patient's life.
Hypergraphia	Keeping extensive diaries, detailed notes; writing autobiography or novel.
Religiosity	Holding deep religious beliefs, often idiosyncratic; multiple conversions, mystical states.
Philosophical interest	Nascent metaphysical or moral speculations, cosmological theories.
Dependence	Cosmic helplessness, "at hands of fate"; protestations of helplessness.
Humorlessness	Overgeneralized ponderous concern; humor lacking or idiosyncratic.
Paranoia	Suspicious, overinterpretation of motives and events; diagnosis of paranoid schizophrenia.

Source: Schomer et al. 2000, 385.

neurophysiologic connections between sensory input and limbic processing (Bear 1979). Ordinary events, therefore, become endowed with extraordinary meaning. A profound metaphysical interpretation, for example, may be developed after a seemingly trivial occurrence. Thus a TLE patient might interpret a routine life event as a sign of divine guidance, and indeed there are descriptions of sudden religious conversions in TLE, sometimes repeatedly occurring in the same patient (Waxman and Geschwind 1975). The role of the temporo-limbic system in religiosity should not be surprising given that this intensely emotional feeling state, like all human experience, is based on neural function

(Saver and Rabin 1997). Just as schizophreniform psychosis is not reversed by elimination of seizures, good seizure control does not appear to influence these personality features. Presumably, subictal irritability is capable of inducing such hyperconnectivity, although the precise connections that are being enhanced is not clear. Pathophysiologically, it has been suggested but not proven that the phenomenon of kindling (Goddard 1983; Meador 2007) may be responsible for this syndrome.

The idea of sensory-limbic hyperconnection gains some additional credence in light of a well-known condition first described in primates and known as the *Klüver-Bucy syndrome* (Klüver and Bucy 1939). These investigators ablated the anterior temporal lobes in male rhesus monkeys and observed a constellation of five changes: hypersexuality, placidity, oral tendencies, visual agnosia or "psychic blindness," and hypermetamorphosis (mandatory environmental exploration). The remarkable disappearance of fear in these animals, to the point where they would calmly approach dangerous snakes that ordinarily would evoke an immediate and strong aversive response, suggests that removal of the amygdalae (LeDoux 2007) was crucial in producing this syndrome. The Klüver-Bucy syndrome has also been described in humans; frontotemporal dementia, in which anterior temporal lobe degeneration can be severe, may show elements of the syndrome or the entire array of clinical features (Cummings and Duchen 1981). Intriguingly, it will be seen that these changes constitute a conceptual opposite to the traits of TLE patients, who have functionally just the opposite lesion—a temporal lobe irritable focus that hyperconnects rather than disconnects the sensory and limbic system (Bear 1979). Hypersexuality in the Klüver-Bucy syndrome, for example, finds its antithesis in the hyposexuality of TLE patients, and placidity is the opposite of emotional intensity. Still, the mechanism whereby a cortical focus of irritability generates abnormalities of information transmission between different areas of the brain remains uncertain. It is also unclear why some patients appear to develop psychosis, whereas others manifest personality change, and still others no significant alterations at all. Advances in the electrographic localization of epileptic areas and in the cellular mechanisms of epileptogenesis will doubtless shed light on these issues.

A particularly troublesome issue is the relationship of TLE to aggression and violence. Violence with the intent to harm or destroy, of course, is a major societal problem, and a deeper understanding of its origins would be helpful not just for improving medical practice but also for advising social policy at national and international levels (Filley et al. 2001). Aggression and violence, like all behaviors, have an anatomy, and legitimate forms of aggression are well recognized, such as defensive aggression. Violence is generally considered a more ominous behavior, implying an unjustified expression of aggression (Filley et al. 2001). Both aggression and violence occur in people with epilepsy, as in all humans, and neurobiological and environmental factors contribute to these behaviors. It is of course true that the premature identification of any neurologic population, such as those with TLE, as more prone to aggression or violence would unjustly stigmatize them. As a general rule, in epilepsy of any type, aggression is uncommon and directed violence during seizures is vanishingly rare (Delgado-Escueta et al. 1981). In TLE, violent automatisms can rarely occur as part of a seizure (Ashford, Schulz, and Walsh 1980), and postictal aggression has also been described (Rodin 1973; Gerard et al. 1998). In these instances, the infliction of harm, if it happens, does not carry any implication of intentionality since the seizure and its aftermath clearly impair consciousness and hence diminish responsibility. Interictal aggression and violence in TLE, however, are more controversial. Case reports of increased interictal aggression in TLE patients have been presented, and animal experiments have shown increased aggression following induction of temporal lobe seizure foci (Devinsky and Vazquez 1993). A number of studies, however, have not found an increase in aggression or violence when TLE patients are compared with those with other types of epilepsy (Hermann and Whitman 1984). Studies of neurosurgical patients have nevertheless noted a high incidence of aggressiveness in TLE patients (Serafetinides 1965), and temporal lobectomy for seizure control may help reduce this behavior (Falconer 1973). In summary, it would appear that whereas aggression is clearly determined by many factors (Pincus and Lewis 1991), some patients with TLE may develop an increased likelihood of interictal aggression and violence. The resolution of this difficult issue must await well-controlled clinical studies (Stevens and Hermann 1981), but recent thinking has tended to minimize the role of temporolimbic systems in aggression and violence while turning more toward the frontal lobes as more crucial in these behaviors (Filley et al. 2001).

The existence of the behavioral syndromes in TLE reviewed in this chapter has been vigorously criticized, and it is true that both the schizophreniform psychosis and the TLE personality suffer from vague definitions and unproven pathogenetic theories. In particular, the controversy over the TLE personality continues, with advocates on both sides of the issue (Devinsky and Najjar 1999; Blumer 1999). While some maintain that chronic subictal cortical discharges alter behavior by kindling the rewiring of temporolimbic circuits dedicated to emotional regulation, others contend that the burdens associated with having epilepsy—whether TLE or any other variety—are enough by themselves to alter personality and behavior without the need to invoke an unproven pathophysiology. A judgmental attitude toward patients with TLE who may be found to have undesirable behavioral traits is of course inappropriate. Rather, the study of these behavioral changes must ultimately seek, as in any medical setting, to help patients in need through the knowledge gained. Moreover, not all potential changes are necessarily undesirable; positive attributes, including artistic creativity and philosophic insight, may also prove to be associated with TLE (Devinsky and Vazquez 1993). What is most intriguing about this often acrimonious debate from a neurobehavioral viewpoint is the possibility of disturbances in sensory-limbic connectivity, an idea that has been and remains amenable to rigorous scientific study. The understanding of brain-behavior relationships demands no less than a willingness to consider and evaluate theoretical behavioral changes due to cerebral disease; TLE is one of several productive research areas that can help achieve this goal.

REFERENCES

Alajouanine, T. Dostoiewski's epilepsy. *Brain* 1963; 86: 209–218.

Ashford, J. W., Schulz, S. C., and Walsh, G. O. Violent automatism in a partial complex seizure. *Arch Neurol* 1980; 37: 120–122.

Bancaud, J., Brunet-Bourgin, F., Chauvel, P., and Halgren, E. Anatomical origin of *déjà vu* and vivid "memories" in human temporal lobe epilepsy. *Brain* 1994; 117: 71–90.

Bard, P. A diencephalic mechanism for the expression of rage with special reference to the sympathetic nervous system. *Am J Physiol* 1928; 84: 490–515.

Bear, D. M. Temporal lobe epilepsy—a syndrome of sensory-limbic hyperconnection. *Cortex* 1979; 15: 357–384.

Bear, D. M., and Fedio, P. Quantitative analysis of interictal behavior in temporal lobe epilepsy. *Arch Neurol* 1977; 34: 454–467.

Benson, D. F. *The Neurology of Thinking*. New York: Oxford University Press; 1994.

Blumer, D. Evidence supporting the temporal lobe epilepsy personality syndrome. *Neurology* 1999; 53 (Suppl 2): S9–S12.

Bogerts, B., Meertz, E., and Schonfeldt-Bausch, R. Basal ganglia and limbic system pathology in schizophrenia: a morphometric study of brain volume and shrinkage. *Arch Gen Psychiatry* 1985; 42: 784–791.

Broca, P. Anatomie comparée de circonvolutions cérébrales: le grand lobe limbique et la scissure limbique dans le série des mammifères. *Rev Anthropol* 1878; 1: 385–498.

Brodal, A. *Neurological Anatomy*. 3rd ed. New York: Oxford University Press; 1981.

Brooks, W. H., Cross, R. J., Roszman, T. L., and Markesbery, W. R. Neuroimmunomodulation: neural anatomical basis for impairment and facilitation. *Ann Neurol* 1982; 12: 56–61.

Carlsson, A. The current status of the dopamine hypothesis of schizophrenia. *Neuropsychopharmacol* 1988; 1: 79–86.

Carpenter, W. T., and Buchanan, R. W. Schizophrenia. *N Engl J Med* 1994; 330: 681–690.

Commission of ILAE, Commission on Classification and Terminology of the International League Against Epilepsy. Proposal for revised classification of epilepsies and epileptic syndromes. *Epilepsia* 1989; 30: 389–399.

Crow, T. J. Molecular pathology of schizophrenia: more than one disease process? *Br Med J* 1980; 280: 66–68.

Cummings, J. L., and Duchen, L. W. Klüver-Bucy syndrome in Pick disease: clinical and pathologic correlations. *Neurology* 1981; 31: 1415–1422.

Currie, S., Heathfield, K.W.G., Henson, R. A., and Scott, D. F. Clinical course and prognosis of temporal lobe epilepsy. *Brain* 1971; 94: 173–190.

Dalgleish, T. The emotional brain. *Nat Rev Neurosci* 2004; 5: 582–589.

Damasio, A. R., and Van Hoesen, G. W. Emotional disturbances associated with focal lesions of the limbic frontal lobe. In: Heilman, K. M., and Satz, P., eds. *Neuropsychology of Human Emotion.* New York: Guilford Press; 1983: 85–110.

Darwin, C. *The Expression of the Emotions in Man and Animals.* London: John Murray; 1872.

Davison, K., and Bagley, C. R. Schizophrenia-like psychoses associated with organic disorders of the central nervous system: a review of the literature. In: Herrington, R. N., ed. Current Problems in Neuropsychiatry. *Br J Psychiatry* Special Publication No. 4, 1969: 113–184.

Delgado-Escueta, A. V., Mattson, R. H., King, L., et al. The nature of aggression during epileptic seizures. *N Engl J Med* 1981; 305: 711–716.

Devinsky, O., and Najjar, S. Evidence against the existence of a temporal lobe epilepsy personality syndrome. *Neurology* 1999: 53 (Suppl 2): S13–S25.

Devinsky, O., and Vazquez, B. Behavioral changes associated with epilepsy. *Neurol Clin* 1993; 11: 127–149.

Falconer, M. A. Reversibility by temporal-lobe resection of the behavioral abnormalities of temporal-lobe epilepsy. *N Engl J Med* 1973; 289: 451–455.

Feinstein, J. S., Rudrauf, D., Khalsa, S. S., et al. Bilateral limbic system destruction in man. *J Clin Exp Neuropsychol* 2010; 32: 88–106.

Filley, C. M., Price, B. H., Nell, V., et al. Toward an understanding of violence: neurobehavioral aspects of unwarranted interpersonal aggression. Aspen Neurobehavioral Conference Consensus Statement. *Neuropsychiatry Neuropsychol Behav Neurol* 2001; 14: 1–14.

Flor-Henry, P. Psychosis and temporal lobe epilepsy: a controlled investigation. *Epilepsia* 1969; 10: 363–395.

Friston, K. J., Liddle, P. F., Frith, C. D., et al. The left temporal region and schizophrenia: a PET study. *Brain* 1992; 115: 367–382.

Garciá-Morales, I., de la Peña Mayor, P., and Kanner, A. M. Psychiatric comorbidities in epilepsy: identification and treatment. *Neurologist* 2008; 14: S15–S25.

Gerard, M. E., Spitz, M. C., Towbin, J. A., and Shantz, D. Subacute postictal aggression. *Neurology* 1998; 50: 384–388.

Geschwind, N. Introduction: psychiatric complications in the epilepsies. *McLean Hosp J* 1977; 10: 6–8.

Gloor, P., Olivier, A., Quesney, L. F., et al. The role of the limbic system in experiential phenomena of temporal lobe epilepsy. *Ann Neurol* 1982; 12: 129–144.

Goddard, G. V. The kindling model of epilepsy. *Trends Neurosci* 1983; 6: 275–279.

Gold, J. M., Hermann, B. P., Randolph, C., et al. Schizophrenia and temporal lobe epilepsy: a neuropsychological analysis. *Arch Gen Psychiatry* 1994; 51: 265–272.

Gold, J. M., and Weinberger, D. R. Cognitive deficits and the neurobiology of schizophrenia. *Curr Opin Neurobiol* 1995; 5: 225–230.

Harrison, P. J. The neuropathology of schizophrenia: a critical review of the data and their interpretation. *Brain* 1999; 122: 593–624.

Harrison, P. J., and Weinberger, D. R. Schizophrenia genes, gene expression, and neuropathology: on the matter of their convergence. *Mol Psychiatry* 2005; 10: 40–68.

Herman, J. P., Ostrander, M. M., Mueller, N. K., and Figueiredo, H. Limbic system mechanisms of stress regulation: hypothalamo-pituitary-adrenocortical axis. *Prog Neuropsychopharmacol Biol Psychiatry* 2005; 29: 1201–1213.

Hermann, B. P., Seidenberg, M., and Jones, J. The neurobehavioral comorbidities of epilepsy: can a natural history be developed? *Lancet Neurology* 2008; 7: 151–160.

Hermann, B. P., and Whitman, S. Behavioral and personality correlates of epilepsy: a review, methodologic critique, and conceptual model. *Psychol Bull* 1984; 95: 451–497.

Hughlings Jackson, J. On a particular variety of epilepsy ("intellectual aura"), one case with symptoms of organic brain disease. *Brain* 1888–1889; 11: 179–207.

Hyde, T. M., and Weinberger, D. R. The brain in schizophrenia. *Semin Neurol* 1990; 10: 276–286.

Iritani, S. Neuropathology of schizophrenia: a mini review. *Neuropathology* 2007; 27: 604–608.

Kanner, A. Psychosis of epilepsy: a neurologist's perspective. *Epilepsy Behav* 2000; 1: 219–227.

Klüver, H., and Bucy, P. C. Preliminary analysis of functions of the temporal lobes in monkeys. *Arch Neurol Psychiatry* 1939; 42: 979–1000.

Kubicki, M., McCarley, R. W., and Shenton, M. E. Evidence for white matter abnormalities in schizophrenia. *Curr Opinion Psychiatry* 2005; 18: 121–134.

Landolt, H. Some clinical electroencephalographic correlates in epileptic psychosis (twilight states). *Electroencephalogr Clin Neurophysiol* 1953; 5: 121.

LeDoux, J. The amygdala. *Curr Biol* 2007; 17: R868–R874.

———. Emotion. In: Mountcastle, V. B., Plum, F., and Geiser, S. R., eds. *Handbook of Physiology*. Section 1: The Nervous System. Bethesda, MD: American Physiological Society; 1987: 419–459.

———. Emotion, memory and the brain. *Sci Am* 1994; 270: 50–57.

MacLean, P. D. Psychosomatic disease and the "visceral brain": recent developments bearing on the Papez theory of emotion. *Psychosomatic Med* 1949; 11: 338–353.

Meador, K. J. The basic science of memory as it applies to epilepsy. *Epilepsia* 2007; 48 (Suppl 9): 23–25.

Mendez, M. F., Grau, R., Doss, R. C., and Taylor, J. L. Schizophrenia in epilepsy: seizure and psychosis variables. *Neurology* 1993; 43: 1073–1077.

Nadkarni, S., Arnedo, V., and Devinsky, O. Psychosis in epilepsy patients. *Epilepsia* 2007; 48 (Suppl 9): 17–19.

Nolte, J. *The Human Brain: An Introduction to Its Functional Anatomy*. 5th ed. St. Louis: Mosby Year Book; 2002.

Palmer, B. W., Dawes, S. E., and Heaton, R. K. What do we know about neuropsychological aspects of schizophrenia? *Neuropsychol Rev* 2009; 19: 365–384.

Papez, J. W. A proposed mechanism of emotion. *Arch Neurol Psychiatry* 1937; 38: 725–743.

Perez, M. M., and Trimble, M. R. Epileptic psychosis—diagnostic comparison with process schizophrenia. *Brit J Psychiatry* 1980; 137: 245–249.

Pincus, J. H., and Lewis, D. O. Episodic violence. *Semin Neurol* 1991; 11: 146–154.

Ramani, V., and Gumnit, R. J. Intensive monitoring of interictal psychosis in epilepsy. *Ann Neurol* 1982; 11: 613–622.

Rapoport, J. L., Addington, A. M., Frangou, S., and Psych, M. R. The neurodevelopmental model of schizophrenia: update 2005. *Mol Psychiatry* 2005; 10: 434–449.

Richter-Levin, G., and Akirav, I. Amygdala-hippocampus dynamic interaction in relation to memory. *Mol Neurobiol* 2000; 22: 11–20.

Roberts, G. W. Schizophrenia: a neuropathological perspective. *Br J Psychiatry* 1991; 158: 8–17.

Rodin, E. A. Psychomotor epilepsy and aggressive behavior. *Arch Gen Psychiatry* 1973; 28: 210–213.

Ross, E. D., Homan, R. W., and Buck, R. Differential hemispheric lateralization of primary and social emotions. *Neuropsychiatry Neuropsychol Behav Neurol* 1994; 7: 1–19.

Saver, J. L., and Rabin, J. The neural substrates of religious experience. *J Neuropsychiatry Clin Neurosci* 1997; 9: 498–510.

Schomer, D. L., O'Connor, M., Spiers, P., et al. Temporolimbic epilepsy and behavior. In: Mesulam, M.-M. *Principles of Behavioral and Cognitive Neurology*. 2nd ed. New York: Oxford University Press; 2000: 373–405.

Serafetinides, E. A. Aggressiveness in temporal lobe epileptics and its relation to cerebral dysfunction and environmental factors. *Epilepsia* 1965; 6: 33–42.

Shenton, M. E., Kikinis, R., Jolesz, F. A., et al. Abnormalities of the left temporal lobe and thought disorder in schizophrenia: a quantitative magnetic resonance imaging study. *N Engl J Med* 1992; 327: 604–612.

Slater, E., and Beard, A. W. The schizophrenia-like psychoses of epilepsy. *Br J Psychiatry* 1963; 109: 95–150.

Stevens, J. R., and Hermann, B. P. Temporal lobe epilepsy, psychopathology, and violence: the state of the evidence. *Neurology* 1981; 31: 1127–1132.

Suddath, R. L., Casanova, M. F., Goldberg, T. E., et al. Temporal lobe pathology in schizophrenia: a quantitative magnetic resonance imaging study. *Am J Psychiatry* 1989; 146: 464–472.

Trimble, M. R. Personality disturbances in epilepsy. *Neurology* 1983; 33: 1332–1334.

Van Os, J., and Kapur, S. Schizophrenia. *Lancet* 2009; 374: 635–645.

Walterfang, M., Wood, S. J., Velakoulis, D., and Pantelis, C. Neuropathological, neurogenetic, and neuroimaging evidence for white matter pathology in schizophrenia. *Neurosci Biobehav Rev* 2006; 30: 918–938.

Waxman, S. G., and Geschwind, N. The interictal behavior syndrome of temporal lobe epilepsy. *Arch Gen Psychiatry* 1975; 32: 1580–1586.

Weinberger, D. R. The pathogenesis of schizophrenia: a neurodevelopmental theory. In: Nasrallah, H. A., and Weinberger, D. R., eds. *Handbook of Schizophrenia*, vol. 1: *The Neurology of Schizophrenia*. New York: Elsevier; 1986: 397–406.

Weinberger, D. R., and Berman, K. F. Speculation on the meaning of cerebral metabolic hypofrontality in schizophrenia. *Schizophr Bull* 1988; 14: 157–168.

Williams, D. The structure of emotions reflected in epileptic experiences. *Brain* 1956; 79: 29–67.

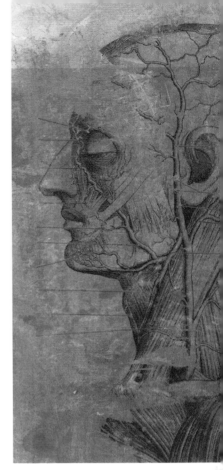

CHAPTER TEN

FRONTAL LOBE
SYNDROMES

The frontal lobes have long fascinated, and perplexed, students of human behavior (Filley 2009). One obvious reason for this is their impressive size, accounting for more than a third of the brain's cortical surface (Damasio and Anderson 2003). Moreover, they are the most phylogenetically recent areas of the brain, and no other animal possesses frontal lobes of such size; accordingly, evolutionary theorists have stressed the role of the frontal lobes in the development of the highest human functions and consciousness itself (Stuss and Benson 1986; Miller and Cummings 1999). It seems certain that the frontal lobes are responsible for singular human capacities, and the period of human evolution has been considered the "age of the frontal lobe" (Tilney 1928). Yet the specific neurobehavioral functions of the frontal lobes remain elusive in many respects. Intimately involved with the highest cognitive and emotional processing, they clearly play a major role in personality, and the maintenance of normal comportment is an essential function (Mesulam 1986). The extensive anatomical connectivity with other regions of the brain (Luria 1980; Damasio and Anderson 2003), however, implies that the frontal lobes participate in all human cognitive and emotional operations, ranging from the simplest to the most evolved behaviors. This powerful organizational influence implies the critical notion of control, and it is instructive to envision the frontal lobes acting as the conductor of the symphony being played by the rest of the brain.

Neuroanatomically located in a position that mirrors their functional importance, the frontal lobes are the most anterior regions of the human brain (Damasio and Anderson 2003). Laterally, they occupy all the area anterior to the Rolandic fissure and superior to the Sylvian fissure, and medially, they extend forward from an imaginary line between the top of the Rolandic fissure and the corpus callosum (Figures 1.1 and 1.2). Numerous parcellations of the frontal lobes have been proposed by neuroanatomists (Stuss and Benson 1986; Miller and Cummings 1999; Damasio and Anderson 2003), but most agree that four functionally distinct regions can be delineated. First, there is the primary motor cortex, Brodmann area 4, well-known to neurologists as the cerebral origin of the corticospinal tract. Second, the premotor area, Brodmann area 6, lies just anterior to the motor cortex and is concerned with the initiation of movement; the medial extension of area 6 is known as the supplementary motor area and plays an important role in the initiation of speech (Chapter 5). Third, Broca's area, corresponding to Brodmann areas 44 and 45 on the left, has a clearly established affiliation with language fluency (Chapter 5), and its analogous zone in the right frontal lobe is thought to subserve language prosody and emotional gesture (Chapter 8). Finally, the remainder of the frontal lobe, Brodmann areas 8–12, 24, 25, 32, 33, 46, and 47, is designated collectively as the prefrontal cortex. In this large and diverse area, the highest functions of the frontal lobes are thought to be represented, and here lesions can produce the various neurobehavioral syndromes considered in this chapter.

As in the case of temporolimbic disorders (Chapter 9), clinicians have long noted that frontal lobe lesions may alter personality (Stuss and Benson 1986; Miller and Cummings 1999; Damasio and Anderson 2003). The concept of *personality* presents difficulties in definition, but we will consider it the characteristic repertoire of behavioral responses used by an individual to relate to the world. Closely related concepts are character and temperament. An individual's personality is determined by a combination of genetic factors and environmental learning, all of which are expressed in the structure and function of the brain (Pinker 2002). Personality is established in the childhood and adolescent years, and, despite the capacity for limited modification later on, the mature adult maintains a stable and lasting personality. On the basis of substantial evidence documenting the alteration of personality and behavior consequent to many disorders affecting the frontal and temporal lobes (Harlow 1868; Damasio and Anderson 2003; Filley and Kleinschmidt-DeMasters 1995; Sollberger et al. 2009), these regions are generally recognized as the cerebral areas where personality is represented. A clear personality change in an adult, therefore, has important neurologic implications, such as the possibility of a frontal lobe meningioma, the onset of frontotemporal dementia, or traumatic brain injury. Personality change in frontal lobe disease or injury is a particularly important clinical problem that offers many opportunities for understanding brain-behavior relationships.

A change in comportment often heralds the onset of frontal lobe involvement and characterizes many aspects of the behavior that is so produced (Mesulam 1986). *Comportment* is a useful concept referring to how a person interacts with others, and this interaction may be disturbed by a breakdown in either cognitive or emotional competence. Since the frontal lobes are linked with both cognitive (neocortical) and emotional (limbic) regions, patients may present with features suggesting cognitive decline—slow thinking, poor judgment, and diminished curiosity—or emotional disorder—inappropriate behavior, social withdrawal, and irritability; often a combination of deficits determines the presentation. The extraordinary range of alterations that can occur testifies to the wide spectrum of mental domains in which the frontal lobes participate.

Many good accounts of personality change after frontal lobe injury have been presented, but the most famous is the extraordinary case of Phineas Gage, a twenty-five-year-old railroad company foreman who sustained a remarkable bifrontal lesion in 1848 (Harlow 1868). While supervising construction of the Rutland and Burlington Railroad across Vermont, Gage was working with a 29-kilogram iron tamping rod in the process of laying an explosive, and during an accidental discharge, the 109-centimeter-long, 3-centimeter-thick rod was propelled at great velocity upward through his face, skull, and brain (Macmillan 1986; Figure 10.1). The tamping rod exited the skull, flew into the sky, and then landed about thirty meters away, where it was later found. Remarkably, he recovered after being momentarily stunned and lived for many years after the injury. With the exception of the loss of his left eye, he appeared to have no physical deficits, and neurologically he had neither aphasia nor paralysis. It was during these years, however, that a dramatic change in personality and comportment was documented (Harlow 1868). Before his accident, he was a responsible, intelligent, and industrious man with a bright future, but after his injury he was noted to become irreverent, profane, and unreliable. He never again showed the promise of his early years, and was clearly "no longer Gage" (Harlow 1868, 340). No autopsy was conducted upon his death at age thirty-six, but his body was later exhumed so that the skull could be recovered; both the skull and the tamping iron are on display at the Warren Anatomical Medical Museum at Harvard University. Computerized reconstruction of the likely trajectory of the tamping rod, based on careful measurements of the damaged skull, have made it clear that the damage involved the right as well as the left frontal lobe, and that the

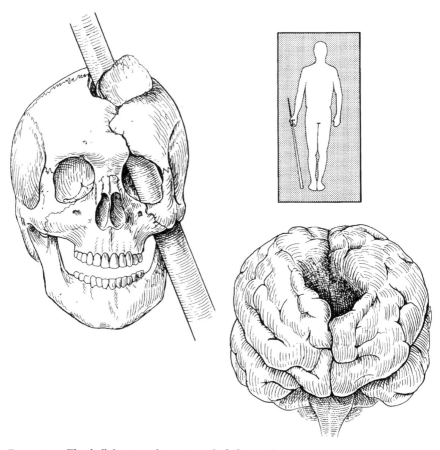

FIGURE **10.1.** *The skull, brain, and tamping rod of Phineas Gage.*

orbitofrontal cortices were most severely affected (Damasio et al. 1994; Figure 10.1). These lesions were therefore the source of the striking behavioral change, as many similar cases of frontal lobe damage since that time have confirmed (Damasio et al. 1994). Like Broca's Leborgne and the amnesic H.M., Phineas Gage demonstrates how the lesion method of behavioral neurology can exploit even a single case to illustrate with singular clarity the relationship of brain and behavior.

The clinical neurologic literature since the time of Phineas Gage has amply confirmed that bilateral frontal lesions are usually necessary for the production of significant neurobehavioral alterations (Mesulam 1986; Damasio and Anderson 2003). Traumatic brain injury, cerebral infarction, neoplasms such as meningioma and glioma, frontotemporal dementia, neurosyphilis, multiple sclerosis, and prefrontal leukotomy are all well-known to cause behavioral disturbances because of damage to both frontal lobes. Unilateral lesions of any etiology are more difficult to identify, probably because of the ability of an intact frontal lobe to compensate for dysfunction in a damaged one. Nevertheless, careful attention to alterations in personality and comportment can uncover many frontal lobe lesions before the onset of most or all routine neurologic symptoms and signs. Although the frontal lobes have been viewed as neurologically "silent" in the sense that they can harbor large lesions without such features as paresis or aphasia appearing, a neurobehavioral approach can substantially assist in the detection of many frontal lesions. The brain is silent only when we lack the skills to listen to it; further study of the frontal lobes will surely lead to refinements in diagnosis based on a deeper knowledge of their unique contributions.

The most characteristic features of the change in personality seen with frontal lobe pathology are impulsive disinhibition and apathetic indifference (Blumer and Benson 1975). Patients can manifest a passive and unmotivated demeanor that unpredictably escalates into euphoric, irritable behavior (Blumer and Benson 1975). Yet this profile fails to embrace many other features patients can demonstrate. It has been pointed out that the neuroanatomic complexity of the frontal lobes and the clinical diversity of frontal lobe dysfunction do not permit a simple delineation of a single frontal lobe syndrome (Damasio and Anderson 2003). We will consider three major syndromes based on areas of primary involvement (Figure 10.2): an *orbitofrontal syndrome*, with disinhibition; a *dorsolateral syndrome*, with executive dysfunction; and a *medial frontal syndrome*, with apathy (Cummings 1993). In practice, elements of these three often coexist since frontal pathology often attains considerable size before the patient comes to clinical attention.

ORBITOFRONTAL SYNDROME

The preeminent feature of the orbitofrontal syndrome is *disinhibition*. The case of Phineas Gage reviewed above clearly exemplifies this disturbance (Harlow 1868; Damasio et al. 1994). Although disinhibition may be strikingly apparent to even casual observation and lead to encounters with the legal system well before a diagnosis can be made, there are no widely accepted quantitative measures of the problem, and careful history-taking and examination offer the most useful means of establishing its presence. Patients with bilateral orbitofrontal (sometimes called ventromedial) lesions engage in inappropriate social, sexual, and dietary behaviors that demonstrate an erosion of the usual restraints by which interpersonal adult life is regulated. Affect may become irritable, labile, euphoric, or unduly jocular. The terms *moria*, meaning an excited affect resembling

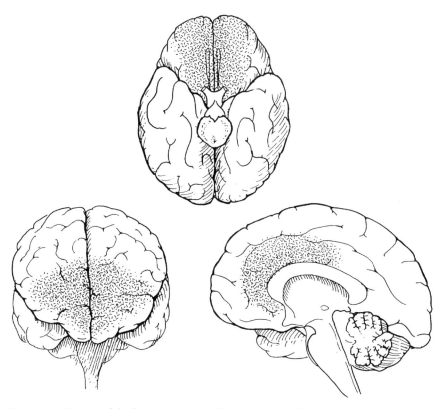

FIGURE 10.2. *Regions of the frontal lobes involved in orbitofrontal, dorsolateral, and medial frontal syndromes.*

hypomania, and *witzelsucht*, the telling of inappropriately caustic or facetious jokes, are both relevant to this syndrome (Hecaen and Albert 1975). Allied with these features is tactlessness, often combined with diminished interpersonal sensitivity and loss of empathy, all contributing to the comportmental dysfunction of these individuals. There is often impaired judgment and insight, such that decisions are made without regard for their consequences (Jarvie 1954; Eslinger and Damasio 1985). Although olfactory function may be affected in this syndrome because of damage to the olfactory nerves at the base of the frontal lobes, elemental neurologic findings may be minimal in such cases, and careful attention to neurobehavioral features is particularly important. Bilateral injury to Brodmann areas 11, 12, and 25 is typically found (Figures 1.3 and 10.2).

The term *acquired sociopathy* has been proposed in the context of this often troublesome syndrome (Eslinger and Damasio 1985). Most problematic is the propensity of such individuals to engage in violent behavior, and recent thinking has evolved to consider frontal damage as more important for the expression of violence than temporal lobe or other neurologic dysfunction (Filley et al. 2001). Many individuals are at risk for violent behavior because of acquired brain damage, often related to traumatic brain injury from motor vehicle accidents or wartime injuries (Chapter 11). In a large study of Vietnam War veterans, ventromedial frontal lobe lesions closely related to orbitofrontal structures were found to increase the risk of violent and aggressive behavior (Grafman et al. 1996). More recently, a critical review of the literature found that orbitofrontal injury was specifically associated with increased aggression (Brower and Price 2001). The relationship of the orbitofrontal syndrome to antisocial personality disorder or sociopathy (American Psychiatric Association 1994) has also drawn attention, as the propensity for violence in people with this personality disorder is well recognized. There are also similarities to the intermittent explosive disorder (American Psychiatric Association 1994). Lesions of the orbitofrontal regions seem to disturb not only the regulation of limbic drives but also the individual's insight into this deficit. It is possible, therefore, that responsibility for one's actions depends upon the integrity of the orbitofrontal cortices and that the disturbing lack of remorse typifying the antisocial personality may reflect dysfunction in this region. In contrast, individuals with intact frontal lobes who have behavioral disinhibition on the basis of the interictal syndrome of TLE appear to have a preserved if not excessive sense of guilt (Chapter 9). A necessary condition for appropriate behavior, then, is the maintenance of a delicate balance between limbic drives and frontal control that integrates the fulfillment of basic needs into the context of responsible action.

These considerations lead us to the recent emergence of the concept of *social cognition* (Frith 2008; Adolphs 2009). In keeping with the nascent enthusiasm of neuroscientists to study emotional topics that were formerly avoided as too subjective, social cognition includes within its scope the capacity of the brain to mediate important interpersonal traits such as empathy, compassion, and altruism. In one intriguing study using a simple gambling task, for example, patients with orbitofrontal lesions did not report regret as a consequence of poor choices, suggesting a diminished capacity for responsible decision making (Camille et al. 2004). Two recent developments have bolstered this line of inquiry, both of which implicate the orbitofrontal regions. First is the idea of *theory of mind*, by which is meant the ability to use one's own beliefs, attitudes, and experiences to understand the mental states of others (Stuss and Anderson 2004). The other is the remarkable discovery of a set of neurons in the brain called *mirror neurons*; these neurons, located in a frontoparietal network including the insula, enable the understanding of others' actions and the intentions behind them (Rizzolatti, Fabbri-Destro, and Cattaneo 2009).

Functional magnetic resonance imaging (fMRI) studies of normal individuals, as could be expected, have identified orbitofrontal regions among a network of areas involved with both theory-of-mind mechanisms and the mirror neuron system (Schulte-Rüther et al. 2007). In view of the apparent role of the orbitofrontal regions not only in impulse control but also in human experiences such as empathy and regret, it is not surprising that damage to these regions has been prominently implicated in violent and criminal behavior (Brower and Price 2001). Injured orbitofrontal cortices have also been found to produce behaviors resembling those of autistic spectrum disorders (Stone, Baron-Cohen, and Knight 1998). Most recently, a laterality effect appears to be emerging whereby the right frontal lobe is selectively responsible for moral behavior (Mendez 2009). Substantial evidence from studies of focal lesions and neurogenerative diseases indicates that the right orbitofrontal region may be particularly crucial for the regulation of social behavior and maintenance of normal comportment (Stuss and Anderson 2004; Rosen et al. 2005; Sollberger et al. 2009).

DORSOLATERAL SYNDROME

In this syndrome, *executive dysfunction* is paramount, implying that the patient is deficient in the planning, sustaining, monitoring, and completion of complex behaviors (Stuss and Benson 1986; Mesulam 1986; Filley 2000). A marked disturbance in the ability to solve problems requiring foresight, goal selection, resistance to interference, use of feedback, and sustained effort is typical. The patient appears inattentive and undermotivated and may make errors on tasks assessing vigilance (Chapter 2). *Perseveration* may be seen (Sandson and Albert 1987), either during the interview or in the course of the examination; failure at an alternating motor sequence task (Chapter 2) can graphically disclose this tendency. *Stimulus-bound behavior* may be encountered, as in the case of a patient who incorrectly draws the hands on a clock at the 10 and the 11 after being asked to indicate the time 11:10. Another phenomenon is echopraxia, the involuntary imitation of the examiner's gestures, which suggests a loss of the internal monitoring of motor behavior (Luria 1973). Associated communication deficits may also be found. First, linguistic impairments related to left-side lesions can occur, including transcortical motor aphasia (Chapter 5) and diminished verbal fluency as measured by word-list generation tasks (Benton 1968). Alternatively, if the lesion is in the right dorsolateral area, transcortical motor aprosody (Chapter 8) can be expected. Damage to Brodmann areas 8, 9, 10, 46, and 47 can be found bilaterally in those who have prominent executive function deficits (Figures 1.3 and 10.2).

In contrast to the orbitofrontal syndrome, neuropsychological deficits are often encountered in this syndrome, but detection of these impairments requires close attention by the neuropsychologist to the process by which a task is approached. A careless and sloppy approach to a standardized test, for example, can indicate prefrontal dysfunction as much as the actual score generated (Stuss and Benson 1986). One measure that has been particularly helpful in the detection of dorsolateral frontal lesions is the Wisconsin Card Sorting Test (WCST), which is designed to assess abstract reasoning and the ability to shift set while avoiding perseveration (Robinson et al. 1980; Filley et al. 1999). The WCST is generally more sensitive to left frontal lesions. In contrast, right frontal lesions may be difficult to identify neuropsychologically, although design fluency tasks—tests probing the facility with which drawings are produced—tend to be maximally impaired with right-side lesions (Jones-Gotman and Milner 1977). Overall, the dorsolateral syndrome is somewhat more identifiable by neuropsychological testing than the orbitofrontal syndrome, and affected patients tend to have a more clearly recognizable cognitive disorder. The primary problem, however, is not in the individual's cognitive capacities but rather in the programming of cognition. The syndrome is characterized by difficulty planning novel cognitive activity and carrying out sequential tasks (Luria 1980). Although cognitive abilities such as memory, language, and visuospatial skills are themselves intact, patients with dorsolateral lesions lack executive control and therefore cannot properly use these skills; in the perceptive words of Luria, "knowledge is divorced from action" (Luria 1973).

MEDIAL FRONTAL SYNDROME

In the medial frontal syndrome, *apathy* is the major characteristic. Affected patients show limited spontaneous movement, gesture, and speech. The term *abulia* refers to more severe forms of apathy. In the most extreme example, immediately following acute bilateral lesions in the medial frontal areas, *akinetic mutism* appears, with the absence of motor activity and speech (Cairns et al. 1941; Ross and Stewart 1981). In akinetic mutism, which can also follow bilateral damage to the midbrain and diencephalon, wakefulness and self-awareness are preserved, but the initiation of behavior is greatly reduced or absent. Transcortical motor aphasia occurs with involvement of the left supplementary motor area (Masdeu, Schoene, and Funkenstein 1978), and the characteristic difficulty in initiating speech is the linguistic manifestation of the general difficulty with movement seen in this syndrome. Lower extremity paresis and gait disturbance can be seen if the lesion extends to the high precentral gyrus. Sphincteric disturbances are common with bilateral medial frontal damage (Andrew and Nathan 1964), and patients with this *frontal lobe incontinence* are characteristically indifferent to the loss of bladder and bowel control. Responsible lesions may be in the anterior cingulate gyri (Neilsen and Jacobs 1951; Barris and Schuman 1953). Brodmann

TABLE **10.1.** Clinical features of frontal lobe syndromes

Orbitofrontal
Disinhibition
Inappropriate affect
Tactlessness
Impaired judgment and insight

Dorsolateral
Executive dysfunction
Perseveration
Stimulus-bound behavior
Impaired word-list generation

Medial frontal
Apathy or abulia
Mutism or transcortical motor aphasia
Lower extremity paresis
Incontinence

areas 24, 32, and 33 are implicated in this syndrome (Figures 1.3 and 10.2).

The medial frontal cortex has been increasingly linked to voluntary action selection (Rushworth 2008). The loss of spontaneity and initiative encompassed by the medial frontal syndrome has led to the provocative but tenuous speculation that the anterior cingulate region may have special importance in the familiar philosophic concept of free will (Crick 1994; Damasio 1994). Although whether human beings even possess free will is subject to philosophical debate, in clinical practice there are patients in whom the capacity for autonomous action seems highly suspect. Patients with the medial frontal syndrome seem to lack the drive to engage in cognitive and emotional life even though the structures subserving the content of this engagement are themselves intact. Thus the behavior implies a diminished or absent will to act. It is, of course, hazardous to assign a notion such as free will to a single area of the brain, and it is doubtful that a discrete "center of the will" exists. However, the passive, even inert, demeanor of patients with medial frontal lesions suggests that a vital component of voluntary action has been compromised.

FUNCTIONS OF THE FRONTAL LOBES

The singular development of the human frontal lobes, particularly the prefrontal regions, strongly implies that they make a unique and critical contribution to mental life. Yet the precise description of this role is exceedingly difficult, and the "riddle of frontal lobe function" (Teuber 1964) remains largely unsolved. The frontal lobes participate in all the higher functions considered in this book, operating in incompletely understood ways to supervise the entire repertoire of human behavior. The key to understanding the highest cerebral functions clearly lies within the expanses of the frontal lobes, but this task is only just beginning (Filley 2009).

It might be imagined that *intelligence* is a higher function that would logically be localized in the frontal lobes, but standard neuropsychological assessments of intelligence—most notably using the intelligence quotient (IQ)—may be normal in patients with significant frontal lobe involvement (Eslinger and Damasio 1985). Intelligence, of course, is a difficult concept to define, and Gould (1981) has vigorously criticized the use of the IQ as a single metric to describe an individual's intellectual capacity. Gardner (1983) has presented an attractive alternative, with his theory of multiple intelligences, that serves to expand the notion of IQ to include a variety of abilities not adequately evaluated by paper-and-pencil tests. Nevertheless, the IQ, as derived from the Wechsler Adult Intelligence Scale–Four (WAIS-4; Wechsler, Coalson, and Raiford 2008), remains a widely used neuropsychological instrument, and some idea of its neurobehavioral interpretation is helpful.

As a general rule, the IQ reflects the function of the brain as a whole, which is in turn dependent on a myriad of genetic and environmental factors influencing neuronal number, synaptic density, myelin integrity, glial cell function, and neurotransmitter metabolism. MRI studies have suggested that brain size, when corrected for body size, does indeed correlate with IQ in normal subjects (Rushton and Ankney 2009). While some evidence can be found for a specific correlation of intelligence and size of the frontal lobes, most neuroimaging studies find that no brain region is particularly associated with intelligence (Rushton and Ankney 2009). The IQ subscores of verbal and performance IQ offer limited insight into left versus right cerebral hemisphere function, but even this distinction is fraught with problems of interpretation (Iverson, Mendrek, and Adams 2004). Thus, as a neuropsychological measure, a person's IQ is best considered as determined by the participation of many cerebral areas, each contributing a unique component based on its functional affiliation. One implication for clinicians is that neither IQ nor, for that matter, any single test reliably identifies frontal lobe lesions (Stuss and Benson 1986; Damasio and Anderson 2003). For theorists of frontal lobe function, these considerations imply that some capacity allied with but superordinate to intelligence lies within these cerebral regions. The frontal lobes seem to have more to do with how a person *uses* intelligence than with intelligence itself.

As discussed above, recent years have also witnessed rapidly expanding knowledge of the role of the frontal lobes in many aspects of emotion and personality known as social cognition (Frith 2008; Adolphs 2009). A central theme is that the frontal lobes serve in a control capacity to supervise other brain regions in the generation of appropriate interpersonal behavior. We have seen in Chapter 2 how the frontal lobes, especially the orbitofrontal regions, are implicated in such positive attributes as empathy (Solberger et al. 2009), altruism (Rilling et al. 2002), and love (Zeki 2007), and these same regions, when dysfunctional, are increasingly identified as contributing to aggression and violence (Filley et al. 2001). In health, the frontal lobes mediate these behaviors for obvious social reasons such as cooperative interaction and social cohesion, and they integrate limbic drives—which may be in some cases, such as defensive aggression, socially and morally justified—into a behavioral repertoire with adaptive value. When the frontal lobes are disturbed, dramatic alterations in emotion and personality may ensue, characterized by loss of empathy, disinhibition, and even violence of a predatory, coldly remorseless nature. What the frontal lobes contribute to the content of consciousness is a balance between limbic drives and socially acceptable comportment.

In a parallel development, and in sharp contrast to the problem of violence, the phenomenon of creativity has also attracted neurobiological attention. Not surprisingly, the frontal lobes also figure prominently in this cherished human capacity (Austin 2003; Heilman 2005). Although the creative mind doubtless gathers input from many neural systems and networks, it is clear that the motivation, executive skills, and interpersonal competence conferred by normal frontal lobes contribute importantly to creativity. Interesting work has also documented the role of the frontal lobes in novelty-seeking (Daffner et al. 2000), raising the fascinating possibility that endogenous, random neural events—likely at the synaptic level—occur in the human brain to generate the novelty underlying creative productivity.

Related to all of these findings, observations have also been made regarding the feeling of happiness as a frontal lobe phenomenon (Berridge and Kringelbach 2008). This work has evolved from the study of reward systems in the brain, which has long been the focus of addiction research, and it appears that a distributed neural network involving the orbitofrontal cortices and the nucleus accumbens, amygdala, and anterior cingulate cortex operates to endow consciousness with the hedonic experience of pleasure (Berridge and Kringelbach 2008). The implications of this knowledge are vast, but as a first step, it is easy to see how the orbitofrontal cortex contributes to a host of core human experiences, both negative—such as the ruthless pleasure a sociopath may derive from inflicting suffering or the euphoric high of the drug addict who exploits others for the next injection—and positive—such as the warmth a person may have from participating in reciprocal altruism or the delight coming from the "ah-hah" moment of creativity.

Notwithstanding the importance of these observations, it is obvious that in many cases we lack the appropriate clinical instruments with which to measure satisfactorily the functions of the frontal lobes. Many impairments relating to the frontal lobes only become apparent in demanding real-life encounters when the individual's capacity to integrate complex information and inhibit inappropriate behaviors is more heavily taxed (Zangwill 1966). Under structured conditions, patients with frontal lobe lesions often respond well to the organization of the setting—the physician's office, the neuropsychology laboratory—and may appear by all criteria quite normal. It is in the everyday world that deficits manifest themselves (Shallice and Burgess 1991); the internal organization provided by the frontal lobes is diminished, and the delicate regulation of cognition and emotions is upset. This aspect of frontal lobe dysfunction further emphasizes the importance of data obtained by a careful clinical history and interview.

Several attempts have been made to capture the essence of frontal lobe function. These are all similar in many respects, and their diversity may only reflect the richness of frontal lobe operations and the different perspectives of the observers. Luria, the influential Russian neuropsychologist, emphasized the hierarchical organization of the brain, at the top of which stand the frontal lobes, which act to regulate "cortical tone" (Luria 1973). Attentional processes figure prominently in this scheme, helping to program and direct mental action. Lhermitte and colleagues have depicted a loss of autonomy in patients with frontal lobe lesions who exhibited inappropriate "imitation" of the examiner's behavior and "utilization" of nearby environmental objects (Lhermitte, Pillon, and Serdaru 1986). This "environmental dependency syndrome" reflects, in Lhermitte's view, a failure to maintain internal control and an excessive dependence on the external environment (Lhermitte 1986). Fuster (1989) proposed that the frontal lobes provide for the temporal structuring of behavior, integrating actions in light of past experience and future plans. Control of interfering stimuli through inhibitory protection is central to his theory. Stuss and Benson (1986) pointed out that the frontal lobes act to regulate all the brain's higher functions. In their view, the frontal lobes mediate self-consciousness and bridge the gap between brain and mind. Mesulam (1986) emphasized the autonomy of behavior provided by the frontal lobes. This feature permits a certain element of freedom from the predictable stimulus-response paradigm of animals with less well-developed frontal lobes. Taking a similar view, Damasio and Anderson (2003) highlighted the guidance of response selection conferred by the frontal lobes. For them, normally functioning frontal lobes offer humans their best chance for long-term survival. These and other notions all have merit, and each contributes to the detailed but still-incomplete picture of how the frontal lobes function. Further study will doubtless clarify the unique supervisory contributions of the frontal lobes, and a single unifying function may emerge as more knowledge is gained.

Perhaps the most useful summary of frontal lobe function can be founded on the neuroanatomic connections of the frontal lobes with sensory, motor, and limbic systems (Nauta 1971; Mesulam 1986; Damasio and Anderson 2003). Interposed between stimulus and response, which in all vertebrate brains are mediated by lower structures including the limbic system (MacLean 1990), the human frontal lobes provide for flexible, autonomous, and goal-directed behavior that considers both past experience and future objectives in the formulation of adaptive behavior (Mesulam 1986; Damasio and Anderson 2003). In the absence of normally functioning frontal lobes, the ephemeral quality of "humanness" is somehow perturbed, despite the preservation of many cognitive and emotional domains that are primarily localized elsewhere in the brain. For the clinician responsible for individuals with frontal lobe disorders, a familiarity with theories of frontal function will prove invaluable in the diagnosis and treatment of patients whose deficits may be as difficult to objectify as they are devastating to effective mental life.

REFERENCES

Adolphs, R. The social brain: the neural basis of social knowledge. *Annu Rev Psychol* 2009; 60: 693–716.

American Psychiatric Association. *Diagnostic and Statistical Manual of Mental Disorders*. 4th ed. Washington, DC: American Psychiatric Association; 1994.

Andrew, J., and Nathan, P. W. Lesions of the anterior frontal lobes and disturbances of micturition and defecation. *Brain* 1964; 87: 233–262.

Austin, J. H. *Chase, Chance, and Creativity*. 2nd ed. Cambridge: MIT Press; 2003.

Barris, R. W., and Schuman, H. R. Bilateral anterior cingulate gyrus lesions. *Neurology* 1953; 3: 44–52.

Benton, A. L. Differential effects in frontal lobe disease. *Neuropsychologia* 1968; 6: 53–60.

Berridge, K. C., and Kringelbach, M. L. Affective neuroscience of pleasure: reward in humans and animals. *Psychopharmacology* 2008; 199: 457–480.

Blumer, D., and Benson, D. F. Personality changes with frontal and temporal lobe lesions. In: Benson D. F., and Blumer, D., eds. *Psychiatric Aspects of Neurologic Disease*. Vol. 1. New York: Grune and Stratton; 1975: 151–169.

Brower, M. C., and Price, B. H. Neuropsychiatry of frontal lobe dysfunction in violent and criminal behavior: a critical review. *J Neurol Neurosurg Psychiatry* 2001; 71: 720–726.

Cairns, J. H., Oldfield, R. C., Pennybacker, J. B., et al. Akinetic mutism with an epidermoid cyst of the 3rd ventricle. *Brain* 1941; 84: 272–290.

Camille, N., Coricelli, G., Sallet, J., et al. The involvement of the orbitofrontal cortex in the experience of regret. *Science* 2004; 304: 1167–1170.

Crick, F. *The Astonishing Hypothesis: The Scientific Search for the Soul*. New York: Charles Scribner's Sons; 1994.

Cummings, J. L. Frontal-subcortical circuits and human behavior. *Arch Neurol* 1993; 50: 873–880.

Daffner, K. R., Mesulam, M.-M., Scinto, L. F., et al. The central role of the prefrontal cortex in directing attention to novel events. *Brain* 2000; 123: 927–939.

Damasio, A. R. *Descartes' Error*. New York: Putnam; 1994.

Damasio, A. R., and Anderson, S. W. The frontal lobes. In: Heilman, K. M., and Valenstein, E., eds. *Clinical Neuropsychology*. 4th ed. New York: Oxford University Press; 2003: 404–446.

Damasio, H., Grabowski, T., Frank, R., et al. The return of Phineas Gage: clues about the brain from the skull of a famous patient. *Science* 1994; 264: 1102–1105.

Eslinger, P. J., and Damasio, A. R. Severe disturbance of higher cognition after bilateral frontal lobe ablation: patient EVR. *Neurology* 1985; 35: 1731–1741.

Filley, C. M. Clinical neurology and executive dysfunction. *Semin Speech Lang* 2000; 21: 95–108.

————. The frontal lobes. In: Boller, F., Finger, S., and Tyler, K. L., eds. *Handbook of Clinical Neurology*. Edinburgh: Elsevier; 2009: 95: 557–570.

Filley, C. M., and Kleinschmidt-DeMasters, B. K. Neurobehavioral presentations of brain neoplasms. *West J Med* 1995; 163:19–25.

Filley, C. M., Price, B. H., Nell, V., et al. Toward an understanding of violence: neurobehavioral aspects of unwarranted interpersonal aggression. Aspen Neurobehavioral Conference Consensus Statement. *Neuropsychiatry Neuropsychol Behav Neurol* 2001; 14: 1–14.

Filley, C. M., Young, D. A., Reardon, M. S., and Wilkening, G. N. Frontal lobe lesions and executive dysfunction in children. *Neuropsychiatry Neuropsychol Behav Neurol* 1999; 12: 156–160.

Frith, C. D. Social cognition. *Phil Trans R Soc B* 2008; 363: 2033–2039.

Fuster, J. M. *The Prefrontal Cortex*. 2nd ed. New York: Raven Press; 1989.

Gardner, H. *Frames of Mind: The Theory of Multiple Intelligences*. New York: Basic Books; 1983.

Gould, S. J. *The Mismeasure of Man*. New York: W. W. Norton; 1981.

Grafman, J., Schwab, K., Warden, D., et al. Frontal lobe injuries, violence, and aggression: a report of the Vietnam Head Injury Study. *Neurology* 1996; 46: 1231–1238.

Harlow, J. M. Recovery from the passage of an iron bar through the head. *Mass Med Soc Publ* 1868; 2: 327–346.

Hecaen, H., and Albert, M. L. Disorders of mental functioning related to frontal lobe pathology. In: Benson, D. F., and Blumer, D., eds. *Psychiatric Aspects of Neurologic Disease*. Vol. 1. New York: Grune and Stratton; 1975: 137–149.

Heilman K. M. *Creativity and the Brain*. New York: Psychology Press; 2005.

Iverson, G. L., Mendrek, A., and Adams, R. L. The persistent belief that VIQ-PIQ splits suggest lateralized brain damage. *Appl Neuropsychol* 2004; 11: 85–90.

Jarvie, H. F. Frontal wounds causing disinhibition. *J Neurol Neurosurg Psychiatry* 1954; 17: 14–32.

Jones-Gotman, M., and Milner, B. Design fluency: the invention of nonsense drawings after focal cortical lesions. *Neuropsychologia* 1977; 15: 653–674.

Lhermitte, F. Human autonomy and the frontal lobes; Part II: Patient behavior in complex and social situations: the "environmental dependency syndrome." *Ann Neurol* 1986; 19: 335–343.

Lhermitte, F., Pillon, B., and Serdaru, M. Human autonomy and the frontal lobes; Part I: Imitation and utilization behavior: a neuropsychological study of 75 patients. *Ann Neurol* 1986; 19: 326–334.

Luria, A. R. *Higher Cortical Functions in Man*. New York: Consultants Bureau; 1980.

————. *The Working Brain: An Introduction to Neuropsychology*. New York: Basic Books; 1973.

MacLean, P. D. *The Triune Brain in Evolution*. New York: Plenum; 1990.

Macmillan, M. B. A wonderful journey through skull and brains: the travels of Mr. Gage's tamping iron. *Brain Cogn* 1986; 5: 67–107.

Masdeu, J. C., Schoene, W. C., and Funkenstein, H. Aphasia following infarction of the left supplementary motor area. *Neurology* 1978; 28: 1220–1223.

Mendez, M. F. The neurobiology of moral behavior: review and neuropsychiatric implications. *CNS Spectr* 2009; 14: 608–620.

Mesulam, M.-M. Frontal cortex and behavior. *Ann Neurol* 1986; 19: 320–325.

Miller, B. L., and Cummings, J. L., eds. *The Human Frontal Lobes*. New York: Guilford Press; 1999.

Nauta, W.J.H. The problem of the frontal lobe: a reinterpretation. *J Psychiat Res* 1971; 8: 167–187.

Neilsen, J. M., and Jacobs, L. L. Bilateral lesions of the anterior cingulate gyri. *Bull LA Neurol Soc* 1951; 16: 231–234.

Pinker, S. *The Blank Slate: The Modern Denial of Human Nature*. New York: Viking; 2002.

Rilling, J. K., Gutman, D. A., Zeh, T. R., et al. A neural basis for social cooperation. *Neuron* 2002; 35: 395–405.

Rizzolatti, G., Fabbri-Destro, M., and Cattaneo, L. Mirror neurons and their clinical relevance. *Nat Clin Pract Neurol* 2009; 5: 24–34.

Robinson, A. L., Heaton, R. K., Lehman, R.A.W., and Stilson, D. W. The utility of the Wisconsin Card Sorting Test in detecting and localizing frontal lobe lesions. *J Cons Clin Psychol* 1980; 48: 605–614.

Rosen, H. J., Allison, S. C., Schauer, G. F., et al. Neuroanatomical correlates of behavioural disorders in dementia. *Brain* 2005; 128: 2612–2625.

Ross, E. D., and Stewart, R. M. Akinetic mutism from hypothalamic damage: successful treatment with dopamine agonists. *Neurology* 1981; 31: 1435–1439.

Rushton, J., and Ankney, C. D. Whole brain size and general mental ability: a review. *Int J Neurosci* 2009; 119: 619–732.

Rushworth, M.F.S. Intention, choice, and the medial prefrontal cortex. *Ann NY Acad Sci* 2008; 1124: 181–207.

Sandson, J., and Albert, M. L. Perseveration in behavioral neurology. *Neurology* 1987; 37: 1736–1741.

Schulte-Rüther, M., Markowitsch, H. J., Fink, G. R., and Piefke, M. Mirror neurons and theory of mind mechanisms involved in face-to-face interactions: a functional magnetic resonance approach to empathy. *J Cogn Neurosci* 2007; 19: 1354–1372.

Shallice, T., and Burgess, P. W. Deficits in strategy application following frontal lobe damage in man. *Brain* 1991; 114: 727–741.

Sollberger, M., Stanley, C. M., Wilson, S. M., et al. Neural basis of interpersonal traits in neurodegenerative diseases. *Neuropsychologia* 2009; 47: 2812–2827.

Stone, V. F., Baron-Cohen, S., and Knight, R. T. Frontal lobe contributions to theory of mind. *J Cogn Neurosci* 1998; 10: 640–656.

Stuss, D. T., and Anderson, V. The frontal lobes and theory of mind: developmental concepts from adult focal lesion research. *Brain Cogn* 2004; 55: 69–83.

Stuss, D. T., and Benson, D. F. *The Frontal Lobes*. New York: Raven Press; 1986.

Teuber, H.-L. The riddle of frontal lobe function in man. In: Warren, J. W., and Akert, K., eds. *The Frontal Granular Cortex and Behavior*. New York: McGraw-Hill; 1964: 410–444.

Tilney, F. *The Brain: From Ape to Man*. New York: Hoeber; 1928.

Wechsler, D., Coalson, D. L., and Raiford, S. E. *Wechsler Adult Intelligence Test: Fourth Edition Technical and Interpretive Manual*. San Antonio: Pearson; 2008.

Zangwill, O. Psychological deficits associated with frontal lobe lesions. *Int J Neurol* 1966; 5: 395–402.

Zeki, S. The neurobiology of love. *FEBS Letters* 2007; 581: 2575–2579.

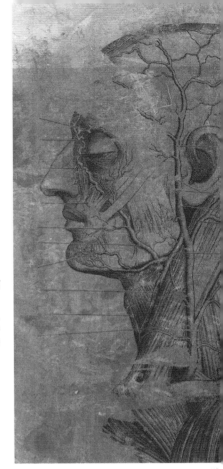

CHAPTER ELEVEN

TRAUMATIC
BRAIN INJURY

Injury to the brain is an obvious but still underemphasized source of neuro-behavioral disability. Although incidence and prevalence figures for *traumatic brain injury* (TBI) are not known with certainty, it is estimated that, in the United States, approximately 1.5 million new cases occur each year (Arciniegas et al. 2005). Of these, some 235,000 require admission to a hospital (Arciniegas et al. 2005), and these patients are of course more easily recognized because of their more severe injuries. Many more TBI victims have so-called mild TBI, which does not require hospital admission but still may produce major clinical problems, often for pro-tracted periods. Motor vehicle accidents, often involving unrestrained or unhel-meted drivers and alcohol abuse; assaults; falls; and sports injuries account for most TBI in civilian populations, and young men aged 15–24 constitute the high-est risk group (Arciniegas et al. 2005). Another peak of incidence occurs in people over 65, who are more vulnerable to the increasingly recognized problem of falls (Arciniegas et al. 2005). Wartime injuries may of course also contribute to a heavy burden of TBI—both in the context of active duty and later on in civilian life—as the recent conflicts in Iraq and Afghanistan make tragically obvious. Penetrating

TBI is well-known to occur from projectile wounds, but the Iraq and Afghanistan wars have brought to light a newly recognized form of TBI known as blast injury (Ling et al. 2009). Overall, TBI ranks as one of the most prevalent neurologic disorders, with an estimated 3.2–5.3 million Americans living with disability subsequent to TBI (Arciniegas et al. 2005; Corrigan, Selassie, and Orman 2010).

Whereas it is true that many individuals with TBI recover uneventfully, survivors are often left with significant and lasting impairments. Most tragic of all are those who remain in the minimally conscious state (MCS; Giacino et al. 2002) or persistent vegetative state (PVS; Multi-Society Task Force on PVS 1994) and never regain any functional independence. Other TBI survivors, who may progress to severe and irreversible dementia, present major problems in social and occupational adjustment. Still other individuals, less obviously impaired because of milder injuries, may be misdiagnosed with psychiatric disorders or not recognized at all. Given the magnitude of cognitive and emotional effects combined with its underdiagnosis, TBI continues to merit its description as a "silent epidemic" (Goldstein 1990).

Closed head injury is the standard term for cranial injury that does not involve a penetrating wound, as would a gunshot injury in time of war. Notwithstanding the important facial, skull, neck, and systemic injuries that can occur and that can preoccupy acute management, this chapter will consider the acute and chronic consequences of closed head injury that are specifically due to TBI. Death can of course occur immediately on impact or during the period before the patient reaches medical attention, and a fatal outcome can also develop later on because of irreversible elevations in intracranial pressure that can occur, particularly in young adults after a second impact within a few days of an injury (Kelly et al. 1991). In those who survive TBI, however, the most disabling long-term consequences are behavioral. A good physical recovery often belies a poor cognitive and especially emotional outcome. The neurobehavioral sequelae of TBI are protean, ranging from postconcussion syndrome to the MCS (Giacino et al. 2002) or PVS (Multi-Society Task Force on PVS 1994) after massive blunt head trauma.

TBI is considered at this point because its clinical diversity provides an appropriate conceptual transition between the focal syndromes reviewed in the last several chapters and the widespread dysfunction represented by dementia, the subject of the concluding chapter. We will follow the useful approach of discussing TBI lesions as either focal or diffuse (Auerbach 1989). Table 11.1 summarizes the neurobehavioral effects of both these varieties.

FOCAL LESIONS

The best-known focal lesions in TBI patients are the intracranial hemorrhages that can easily be detected by computed tomography (CT) scans, often in hospi-

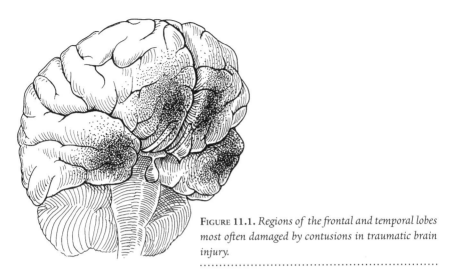

FIGURE 11.1. *Regions of the frontal and temporal lobes most often damaged by contusions in traumatic brain injury.*

tal emergency rooms, and that immediately prompt neurosurgical consultation: subdural hematoma, epidural hematoma, intraparenchymal hematoma, and subarachnoid hemorrhage. These will be discussed below, but for neurobehavioral purposes, the most important focal lesion is the *contusion*, or bruise of the cortical surface (Courville 1937). As the skull is deformed or fractured at the moment of impact, the underlying cortex is damaged, and the injury may also involve subjacent white matter. Typically there is focal hemorrhage and edema along with the neuronal loss and demyelination. Contusions can occur anywhere in the hemispheres but are most likely in the orbitofrontal, frontopolar, and anterior temporal regions, where the brain is thrust forcefully against bony prominences in the anterior and middle cranial fossae (Courville 1937). Figure 11.1 illustrates the distribution of contusions that are likely to result from TBI.

Contusions are classically thought to be found either in a region directly underlying the site of an external blow (the *coup* lesion) or on the other side of the brain (the *contrecoup* lesion). Coup lesions are due to direct contact with the inner surface of the skull, but contrecoup lesions are due to horizontal movement of the brain in an immobile skull that produces a cavitation effect in a region roughly opposite the point of impact. Multiple contusions frequently occur, and it is generally the case that prognosis worsens as the number of contusions increases. By virtue of their propensity to damage frontal and temporal lobes, contusions are strongly associated with disorders of comportment, impulse control, language, and memory (Blumer and Benson 1975; Alexander 1982). Since contusions are usually bilateral (Adams et al. 1980), a single neurobehavioral syndrome is seldom encountered.

Other, more obvious focal lesions are those due to various types of *intra-cranial hemorrhage*. The most important of these for our purposes is *intracerebral hemorrhage*, often in the thalamus or basal ganglia, which can cause parenchymal destruction related to shearing injury of small deep blood vessels (Adams et al. 1986). This kind of hemorrhage implies a poor prognosis, and survivors can show residual aphasia, neglect, or other syndromes. Subdural and epidural hematomas are also common, from bleeding originating in damaged vessels outside the brain itself but within the skull; *subdural hematoma* results from injury to bridging veins in the subdural space, and *epidural hematoma* typically follows rupture of the middle meningeal artery related to an overlying skull fracture. Although common and life-threatening in TBI, these extraparenchymal hematomas are rather unlikely to eventuate in chronic neurobehavioral sequelae if there is no underlying contusion and they are treated successfully (Levin 1992). Similarly, *subarachnoid hemorrhage*, which is common in TBI, usually leaves no sequelae if appropriately treated. In contrast, this kind of bleeding from a ruptured aneurysm is associated with high mortality and morbidity.

Finally, there are two focal lesions that occur in the context of brain herniation, itself a poor prognostic sign because of the potential for destruction of vital centers in the lower brainstem. *Posterior cerebral artery occlusion* resulting from downward herniation of the brain can lead to visual defects such as hemianopia and alexia because of damage to the medial occipital lobe, and, if the occlusion is bilateral, cortical blindness or visual agnosia can result (Alexander 1982). Small, so-called *Duret hemorrhages* in the midline of the brainstem, usually multiple, are due to compression of small arteries and veins in the course of rostral-caudal herniation and contribute to stupor and coma (Alexander 1982).

DIFFUSE LESIONS

The destructive effects of focal lesions associated with TBI are often severe, but still more disabling are the sequelae of diffuse lesions. A wide spectrum of syndromes can follow diffuse trauma to the brain, ranging from concussion with self-limited attentional and mood disorders to irreversible dementia, MCS, and PVS. The most common cause of PVS is in fact TBI, and next in importance is hypoxic-ischemic injury (see below), which can often be associated with trauma (Multi-Society Task Force on PVS 1994). This problem alone raises many difficult medical and ethical issues, but PVS is only one of the many chronic disorders associated with diffuse TBI lesions.

Many patients with TBI are seen when clinical or neuroradiologic evidence of brain swelling is present. Paradoxically, this alarming acute effect is frequently transient and without aftermath. Although it would seem that brain swelling, with its attendant risk of herniation, would be a frequent source of neurobehavioral dysfunction in TBI, there is usually little residual effect if treatment is prompt and effective. *Brain swelling* is a nonspecific term for increased brain volume due to an increase in water content, which in TBI can be from vaso-genic edema (Fishman 1975), cytotoxic edema associated with hypoxic-ischemic injury (Fishman 1975), or cerebrovascular congestion (Kelly et al. 1991). Swelling of the brain does not itself cause injury, and if herniation is avoided, no chronic sequelae of the transiently increased brain volume are evident (Bruce et al. 1981; Auerbach 1989). When herniation occurs, of course, the outcome may be cata-strophic, with compression of the diencephalon and brainstem leading to death. In terms of chronic neurobehavioral sequelae, however, other diffuse lesions are highly significant.

The diffuse lesions that occur in the chronic phase after TBI have been the subject of some controversy over the past several decades, but recently a consensus has been reached that the entity of *diffuse axonal injury* (DAI) is a frequent and clinically significant neuropathologic finding in blunt TBI. The term traumatic axonal injury is an alternative name for this lesion. Interestingly, penetrating TBI is infrequently associated with DAI, since the wound usually involves focal damage without widespread injury. First described in the 1950s as "diffuse degeneration of the cerebral white matter" (Strich 1956), damage affecting axons and myelin after TBI has been documented by many clinical observations and experimental studies (Oppenheimer 1968; Adams et al. 1982; Gennarelli et al. 1982; Smith, Meaney, and Shull 2003; Vannorsdall et al. 2010). These lesions are widespread and multifocal, involving the brainstem, corpus callosum, and hemispheric white matter, and are related to the biomechanics of shearing injury in the white matter consequent to acceleration and decel-eration (Gennarelli et al. 1982). DAI is more likely to be caused by rotational (angular) forces, such as those that occur during motor vehicle accidents, than by translational (linear) forces (Adams et al. 1982). The lesions of DAI may also be caused by acceleration-deceleration injury without any impact to the head (Gennarelli et al. 1982). The characteristic microscopic features of DAI include shearing injury to axons, with cytoskeletal damage and retraction balls reflect-ing secondary axotomy, and blood vessel injury resulting in microhemorrhages (Smith, Meaney, and Shull 2003). It has also become apparent that a host of injury-mediated cytotoxic processes evolve over days, weeks, and even months after the injury, involving excess intracellular sodium and calcium, free radi-cal formation, and damage from inflammatory cytokines (Smith, Meaney, and Shull 2003; Arciniegas et al. 2005). Figure 11.2 presents a schematic view of a brain damaged by DAI.

One of the problems in the clinical investigation of TBI, however, is that diagnosis of mild forms remains inexact. Whereas a moderately or severely injured patient is readily recognized by clinical and often neuroradiologic obser-

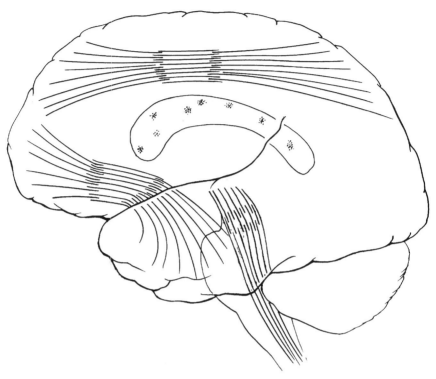

FIGURE 11.2. *Disruption of white matter in the brainstem, cerebral hemispheres, and corpus callosum as a result of diffuse axonal injury.*

vations, people with mild TBI, especially days or weeks later, may appear normal at first glance, and if the details of the TBI event are not available, symptoms may be ascribed to psychological or other factors. Despite improvements in CT and magnetic resonance imaging (MRI), it is still true that the great majority of mild TBI patients show no abnormality on conventional structural imaging (Lee and Newberg 2005). Gradient echo MRI sometimes shows small petechial hemorrhages in the white matter reflective of areas of DAI, but this is not reliable in the usual patient. Despite much interest in positron emission tomography (PET), single photon emission computed tomography (SPECT), and functional MRI (fMRI; Belanger et al. 2007), these modalities cannot be used as diagnostic tests for TBI; a thorough analysis of SPECT for this purpose, for example, showed significant inadequacies (Wortzel et al. 2008). Of all the newer imaging techniques, the evolving technique of diffusion tensor imaging (DTI) has the most promise in TBI because it allows detailed imaging of white matter areas expected to be injured as a result of DAI (Kraus et al. 2007; Sidaros et al. 2008).

Recent DTI findings in mild TBI have in fact documented frontal white matter disturbances that correlate with impaired executive function (Lipton et al. 2009). Yet still, clinical criteria remain most helpful for diagnostic and prognostic purposes in TBI.

The duration of unconsciousness after TBI is an important clinical datum that has been correlated with outcome (Gilchrist and Wilkinson 1979). TBI causing loss of consciousness does so immediately upon impact, and DAI involving the dorsolateral midbrain and pons is responsible for post-traumatic coma (Adams et al. 1982; Gennarelli et al. 1982). The midbrain is particularly vulnerable to rotational injury, and the ascending reticular activating system (ARAS) in the brainstem that mediates arousal is thereby disrupted (Chapter 3). As might be expected, the degree of brainstem damage from DAI reflects the severity of the injury, and the longer the period of coma, the poorer the outcome. In contrast, mild degrees of DAI with less brainstem damage result in a better outcome (Levin et al. 1988). DAI can also severely disrupt the hemispheric white matter, and in cases of PVS that follow TBI, the hemispheres exhibit more severe DAI than does the brainstem (Multi-Society Task Force on PVS 1994). Thus DAI is present in TBI cases representing a wide range of severity and is consistently associated with neurobehavioral manifestations.

Also important in terms of predicting outcome is a phase of recovery known as post-traumatic amnesia (PTA), the period after the recovery of consciousness before a patient regains the ability to acquire new learning. In contrast to the duration of unconsciousness, which is frequently unavailable or uncertain, the length of PTA can be more useful since it is often observable within a medical setting and can be accurately determined. Studies of TBI patients have revealed that the longer the duration of PTA, the greater the degree of cognitive disability (Brooks et al. 1980). The neuropathological basis of PTA is uncertain, but DAI in temporal lobe and limbic systems mediating memory function seems likely.

In severe TBI, the most disruptive changes that attend the chronic recovery state are in the realm of personality (Bleiberg, Cope, and Spector 1989). Disinhibition, apathy, and deterioration in comportment may all occur, and the severity of these changes generally correlates with the duration of coma (Levin and Grossman 1978). These features are reminiscent of deficits seen with frontal lobe involvement (Chapter 10), and it is of interest to consider the neuropathologic basis of these alterations. Although frontal lobe contusions would be one apparent mechanism leading to the development of frontal lobe behavioral features, DAI may in fact be responsible more often. In the chronic phase following TBI, ventricular enlargement associated with DAI results in a poor neurobehavioral outcome (Filley et al. 1987; Vannorsdall et al. 2010), implying that frontal-limbic white matter connections have been disrupted. We shall discuss the importance of cerebral white matter more extensively in Chapter 12.

In recent years, the problem of mild head injury has received increased attention as a source of considerable suffering and disability (Alexander 1995; McAllister and Arciniegas 2002; Randolph et al. 2009). Difficulties with defining this entity, as well as the equally obscure term minor head injury, have beclouded this area, but it has long been clear that significant neurobehavioral disability can follow seemingly trivial head injuries (Rimel et al. 1981). A more precise term is concussion, which has traditionally been defined as the loss of consciousness resulting from a blow to the head (Geisler and Greenberg 1990). However, it is clear that TBI can occur without loss of consciousness and may be manifest by confusion or amnesia alone (Fisher 1966; Kelly et al. 1991; Kelly 1999). These observations suggest that the deep midline arousal structures of the ARAS can be unaffected while more superficial brain regions can be damaged. This notion is supported by data obtained from experimental animals indicating that TBI affects the most peripheral cerebral regions before it damages the brainstem (Ommaya and Gennarelli 1974). It therefore appears that a more appropriate definition of *concussion* is a traumatically induced alteration in mental status (Kelly et al. 1991; Kelly 1999). With the use of this definition, concussion serves well as a single term encompassing all degrees of mild head injury and provides a solid basis for further clinical and theoretical considerations regarding TBI.

Concussion has classically been taught to be reversible, in that no long-term sequelae of any kind result. However, the *postconcussion syndrome* (PCS) has recently become well-known as a frequent cause of persistent headache, inattention, memory disturbance, fatigue, dizziness, insomnia, anxiety, and depression. Although most individuals experience a rapid and complete recovery (Alexander 1995), the PCS can often be troublesome for weeks, months, or even longer after the injury. The usefulness of considering this syndrome as a uniform entity has been questioned (McAllister and Arciniegas 2002), and undoubtedly psychogenic and litigation issues often contribute to symptoms, especially as the duration of the PCS extends. However, for our purposes, it appears that the most consistent lesion in concussion, as in severe TBI, is DAI (Alexander 1995; Smith, Meaney, and Shull 2003). Since few persons die with recent concussions, autopsied cases are rare, but available postmortem human observations (Oppenheimer 1968; Bigler 2004) and experimental studies (Ommaya and Gennarelli 1974; Povlishock 1992) suggest that DAI is the underlying lesion. In parallel with this common neuropathology is the qualitative clinical similarity of concussion and severe TBI; in both syndromes, disruption of brainstem and hemispheric white matter tracts creates persistent problems in arousal, attention, concentration, and mood (Alexander 1995). DAI is the common denominator of TBI, and the degree of injury determines the severity but not the essential quality of the clinical picture (Alexander 1995).

As mentioned above, the recent Middle East conflicts have introduced a new form of diffuse injury related to TBI. *Blast injury* is a form of TBI that follows exposure to a rapid burst of hot air, often commingled with debris and shrapnel, originating from a nearby explosion (Ling et al. 2009). A sudden high-pressure incident impacts the head and results in damage to all intracranial contents in a relatively uniform manner; one observation in support of this hypothesis is that subarachnoid hemorrhage has often been noted, implying an effect on vascular as well as neural structures (Ling et al. 2008). However, the pathophysiology of blast injury is not known, as experimental and neuropathological studies are lacking. Whether blast injury differs substantially from standard TBI is not yet determined (Ling et al. 2009), but both surely injure the brain in a diffuse manner, and until more data are available, it is reasonable to consider the two as similar for clinical purposes.

Another major category of diffuse lesion is *hypoxic-ischemic injury* (Alexander 1982). Many TBI cases are complicated by systemic injury leading to hypoxia or hypotension. These insults may occur in up to 90 percent of severe TBI cases and have primary effects on particularly oxygen-dependent areas such as the hippocampus, basal ganglia, and cerebral cortex (Graham, Adams, and Doyle 1978). Watershed areas of the cortex are especially vulnerable to hypotensive events. In general, superimposed hypoxic-ischemic injury worsens the prognosis of TBI and often contributes to severe dementia, MCS, or PVS (Graham et al. 1983; Giacino et al. 2002; Multi-Society Task Force on PVS 1994).

The final category of diffuse lesions in TBI is the syndrome of *chronic traumatic encephalopathy* (CTE) in athletes exposed to repeated sublethal TBI (McKee et al. 2009). The problem of sports concussion has received much recent attention, as it is estimated that in the United States, 300,000 cases of TBI occur each year in the context of sports and recreation (Kelly 1999). Boxing, however, is the most obvious setting in which TBI is encountered. Although athletes engaged in other contact sports have been observed to develop CTE, boxers are likely to have the highest risk of this progressive dementia since the objective of boxing is nothing less than the production of TBI. CTE recalls an older literature in which the term *dementia pugilistica* was introduced in the 1930s (Millspaugh 1937) as an alternative to the colorful but derisive "punch drunk encephalopathy" (Martland 1928). This disorder was reported as a dementing illness in boxers, characterized by dementia, parkinsonism, and ataxia (Martland 1928; Millspaugh 1937), and these clinical features are quite similar to the dementia, parkinsonism, and speech and gait disorders seen in CTE. Boxers have also been studied neuropsychologically, and cognitive deficits have been documented in both active (Drew et al. 1986) and retired (Casson, Siegel, and Sham 1984) contestants, implying that brain injury occurs early in a fighter's career and progresses later (Unterharnscheidt 1970). The severity of cognitive impairment appears to be related to the duration of the pugilist's career and the number of blows received (Unterharnscheidt 1970). CT scans of boxers often reveal brain atrophy, and a

TABLE 11.1. TBI lesions and their neurobehavioral effects

	Lesion	Site(s)	Sequelae
Focal	Contusion	Frontal lobes Temporal lobes	Disinhibition, apathy, amnesia, aphasia
	Intracerebral hemorrhage	Thalamus Basal ganglia	Initial coma; residual aphasia, hemineglect
	Posterior cerebral artery occlusion	Occipital lobe	Hemianopia, blindness, alexia, visual agnosia
	Duret hemorrhages	Brainstem	Progressive stupor and coma
Diffuse	Diffuse axonal injury	Brainstem Cerebrum Corpus callosum	Initial coma, amnesia; residual MCS or PVS, inattention, personality change
	Blast injury	Brainstem Hippocampus	Initial coma, amnesia
	Hypoxic-ischemic injury	Hippocampus Neocortex	Amnesia, dementia, MCS, PVS
	Chronic traumatic encephalopathy	Neocortex Hippocampus Substantia nigra Cerebellum	Dementia, parkinsonism, ataxia

MCS: minimally conscious state; PVS: persistent vegetative state

cavum septum pellucidum—a fluid-filled cavity within the membrane separating the lateral ventricles called the septum pellucidum—is sometimes seen as well (Casson et al. 1982).

Neuropathological study of CTE and dementia pugilistica has shown diffuse atrophy and abundant neurofibrillary tangles in the neocortex, hippocampus, limbic system, substantia nigra, white matter, brainstem, and cerebellum; notably, there are relatively few neuritic plaques (Corsellis, Bruton, and Freeman-Browne 1973; McKee et al. 2009). The neuropathological changes account well for the observed cognitive and motor features. The prominence of neurofibrillary tangles has led to speculation regarding the possibility of TBI being a risk factor for Alzheimer's disease (AD; Chapter 12). Indeed, epidemiologic evidence suggests that such an association is likely based on a synergistic effect of the apolipoprotein E ε4 (APOE4) allele with TBI in the development of AD (Mayeux et al. 1993; Mauri et al. 2006). However, the paucity of neuritic plaques in CTE and dementia pugilistica differs distinctly from the neuropathology of AD, and the constellation of neuropathological findings favors the view that tau pathology is a consequence of repeated TBI in the boxing ring or other athletic settings (Corsellis, Bruton, and Freeman-Browne 1973; McKee et al. 2009). Axonal injury is present in CTE, and the deposition of tau protein in white matter is suspected to arise from axonal damage (McKee et al. 2009), consistent with the notion that DAI is one of the contributing factors in the pathogenesis of CTE. Taken together, the data suggest that TBI in early life may predispose to the later onset of two different progressive dementias: the tauopathy of CTE and typical AD in those who harbor the APOE4 allele.

It is thus inescapable that permanent damage to the brain may occur as a result of boxing, and often with other contact sports as well. In boxing, of course, a knockout means nothing less than a blow of sufficient force to produce immediate coma, and lasting brain injury is known to result even from repetitive blows that do not involve loss of consciousness (Lampert and Hardman 1984). From a preventive perspective, it is noteworthy that neither the boxer's skill as a pugilist (Casson, Siegel, and Sham 1984) nor the introduction of medical supervision and safety measures (Drew et al. 1986) appears to prevent the onset of cognitive decline. This "noble art of self-defense" therefore represents the deliberate and sanctioned infliction of injury to the brain—injury that society otherwise assiduously strives to avoid. Other contact sports also involve substantial risk of TBI, and although the objective in these contests is not the infliction of brain injury, the hazards of severe neurobehavioral impairment or even, in rare cases, death (Kelly et al. 1991) are sobering. These emerging neurologic risks, which may lead to a range of devastating effects acutely and over an athlete's lifetime, have prompted a reconsideration of the safety of contact sports and renewed interest in effective prevention of TBI among athletes (Kelly 1999).

REFERENCES

Adams, J. H., Doyle, D., Graham, D. I., et al. Deep intracerebral (basal ganglia) hematomas in fatal non-missile head injury in man. *J Neurol Neurosurg Psychiatry* 1986; 49: 1039–1043.

Adams, J. H., Graham, D. I., Murray, L. S., and Scott, G. Diffuse axonal injury due to non-missile head injury: an analysis of 45 cases. *Ann Neurol* 1982; 12: 557–563.

Adams, J. H., Scott, G., Parker, L. S., et al. The contusion index: a quantitative approach to cerebral contusions in head injury. *Neuropathol Appl Neurobiol* 1980; 6: 319–324.

Alexander, M. P. Mild traumatic brain injury: pathophysiology, natural history, and clinical management. *Neurology* 1995; 45: 1252–1260.

———. Traumatic brain injury. In: Benson D. F., and Blumer, D., eds. *Psychiatric Aspects of Neurologic Disease.* Vol. 2. New York: Grune and Stratton; 1982: 219–248.

Arciniegas, D. B., Anderson, C. A., Topkoff, J., and McAllister, T. W. Mild traumatic brain injury: a neuropsychiatric approach to diagnosis, evaluation, and treatment. *Neuropsychiatric Dis Treat* 2005; 1: 311–327.

Auerbach, S. H. The pathophysiology of traumatic brain injury. In: Horn, L. J., and Cope, D. N., eds. *Traumatic Brain Injury.* Philadelphia: Hanley and Belfus; 1989: 1–11.

Belanger, H. G., Vanderploeg, R. D., Curtiss, G., and Warden, D. L. Recent neuroimaging techniques in mild traumatic brain injury. *J Neuropsychiatry Clin Neurosci* 2007: 19: 5–20.

Bigler, E. D. Neuropsychological results and neuropathological findings at autopsy in a case of mild traumatic brain injury. *J Int Neuropsychol Soc* 2004; 10: 794–806.

Bleiberg, J., Cope, D. N., and Spector, J. Cognitive assessment and therapy in traumatic brain injury. In: Horn, L. J., and Cope, D. N., eds. *Traumatic Brain Injury*. Philadelphia: Hanley and Belfus; 1989: 95–121.

Blumer, D., and Benson, D. F. Personality changes with frontal and temporal lobe lesions. In: Benson, D. F., and Blumer, D., eds. *Psychiatric Aspects of Neurologic Disease*. Vol. 1. New York: Grune and Stratton; 1975: 151–169.

Brooks, D. N., Aughton, M. E., Bond, M. R., et al. Cognitive sequelae in relationship to early indices of severity of brain damage after severe blunt head injury. *J Neurol Neurosurg Psychiatry* 1980; 43: 529–534.

Bruce, D. A., Alavi, A., Bilaniuk, L., et al. Diffuse cerebral swelling following head injuries in children: the syndrome of "malignant brain edema." *J Neurosurg* 1981; 54: 170–178.

Casson, I. R., Sham, R., Campbell, E. A., et al. Neurological and CT evaluation of knocked-out boxers. *J Neurol Neurosurg Psychiatry* 1982; 45: 170–174.

Casson, I. R., Siegel, O., and Sham, R. Brain damage in modern boxers. *J Am Med Assoc* 1984; 251: 2663–2667.

Corrigan, J. D., Selassie, A. W., and Orman, J. A. The epidemiology of traumatic brain injury. *J Head Trauma Rehabil* 2010; 25: 72–80.

Corsellis, J.A.N., Bruton, C. J., and Freeman-Browne, D. The aftermath of boxing. *Psychol Med* 1973; 3: 270–303.

Courville, C. B. *Pathology of the Central Nervous System*. Mountain View, CA: Pacific; 1937.

Drew, R. H., Templer, D. I., Schuyler, B. A., et al. Neuropsychological deficits in active licensed professional boxers. *J Clin Psychol* 1986; 42: 520–525.

Filley, C. M., Cranberg, L. D., Alexander, M. P., and Hart, E. J. Neurobehavioral outcome after closed head injury in childhood and adolescence. *Arch Neurol* 1987; 44: 194–198.

Fisher, C. M. Concussion amnesia. *Neurology* 1966; 16: 826–830.

Fishman, R. A. Brain edema. *N Engl J Med* 1975; 293: 706–711.

Geisler, F. H., and Greenberg, J. Management of the acute head-injury patient. In: Salcman, M., ed. *Neurologic Emergencies*. 2nd ed. New York: Raven Press; 1990: 135–165.

Gennarelli, T. A., Thibault, L. E., Adams, J. H., et al. Diffuse axonal injury and traumatic coma in the primate. *Ann Neurol* 1982; 12: 564–574.

Giacino, J. T., Ashwal, S., Childs, N., et al. The minimally conscious state: definition and diagnostic criteria. *Neurology* 2002; 58: 349–353.

Gilchrist, E., and Wilkinson, M. Some factors determining prognosis in young people with severe head injuries. *Arch Neurol* 1979; 36: 355–359.

Goldstein, M. Traumatic brain injury: a silent epidemic. *Ann Neurol* 1990; 27: 327.

Graham, D. I., Adams, J. H., and Doyle, D. Ischemic brain damage in fatal non-missile head injuries. *J Neurol Sci* 1978; 39: 213–234.

Graham, D. I., McClellan, D., Adams, J. H., et al. The neuropathology of severe disability after head injury. *Acta Neurochir* (Suppl) 1983; 32: 65–67.

Kelly, J. P. Traumatic brain injury and concussion in sports. *J Am Med Assoc* 1999; 282: 989–991.

Kelly, J. P., Nichols, J. S., Filley, C. M., et al. Concussion in sports: guidelines for the prevention of catastrophic outcome. *J Am Med Assoc* 1991; 266: 2867–2869.

Kraus, M. F., Susmaras, T., Caughlin, B. P., et al. White matter injury and cognition in chronic traumatic brain injury: a diffusion tensor imaging study. *Brain* 2007; 130: 2508–2519.

Lampert, P. W., and Hardman, J. M. Morphological changes in brains of boxers. *J Am Med Assoc* 1984; 251: 2676–2679.

Lee, B., and Newberg, A. Neuroimaging in traumatic brain injury. *NeuroRx* 2005; 2: 372–383.

Levin, H. S. Neurobehavioral recovery. *J Neurotrauma* 1992; 9: S359–S373.

Levin, H. S., and Grossman, R. G. Behavioral sequelae of closed head injury. *Arch Neurol* 1978; 35: 720–727.

Levin, H. S., Williams, D., Crofford, M. J., et al. Relationship of depth of brain lesions to consciousness and outcome after closed head injury. *J Neurosurg* 1988; 69: 861–866.

Ling, G., Bandak, F., Armonda, R., et al. Explosive blast neurotrauma. *J Neurotrauma* 2009; 26: 815–825.

Lipton, M. L., Gulko, E., Zimmerman, M. E., et al. Diffusion-tensor imaging implicates prefrontal axonal injury in executive function impairment following very mild traumatic brain injury. *Radiology* 2009; 252: 816–824.

Martland, H. S. Punch drunk. *J Am Med Assoc* 1928; 91: 1103–1107.

Mauri, M., Sinforian, E., Bono, G., et al. Interaction between apolipoprotein epsilon 4 and traumatic brain injury in patients with Alzheimer's disease and mild cognitive impairment. *Funct Neurol* 2006; 21: 223–228.

Mayeux, R., Ottman, R., Tang M.-X., et al. Genetic susceptibility and head injury as risk factors for Alzheimer's disease among community-dwelling elderly persons and their first-degree relatives. *Ann Neurol* 1993; 33: 494–501.

McAllister, T. W., and Arciniegas, D. Evaluation and treatment of postconcussive symptoms. *NeuroRehabilitation* 2002; 17: 265–283.

McKee, A. C., Cantu, R. C., Nowinski, C. J., et al. Chronic traumatic encephalopathy in athletes: progressive tauopathy after repetitive head injury. *J Neuropathol Exp Neurol* 2009; 68: 709–735.

Millspaugh, J. A. Dementia pugilistica (punch drunk). *US Nav Med Bull* 1937; 35: 297–303.

Multi-Society Task Force on PVS. Medical aspects of the persistent vegetative state. *N Engl J Med* 1994; 330: 1499–1508, 1572–1579.

Ommaya, A. K., and Gennarelli, T. A. Cerebral concussion and traumatic unconsciousness. *Brain* 1974; 97: 633–654.

Oppenheimer, D. R. Microscopic lesions in the brain following head injury. *J Neurol Neurosurg Psychiatry* 1968; 31: 299–306.

Povlishock, J. T. Traumatically induced axonal injury: pathogenesis and pathobiological implications. *Brain Pathol* 1992; 2: 1–12.

Randolph, C., Millis, S., Barr, W. B., et al. Concussion symptom inventory: an empiri-
cally derived scale for monitoring resolution of symptoms following sport-related
concussion. *Arch Clin Neuropsychol* 2009; 24: 219–229.

Rimel, R. W., Giordani, B., Barth, J. T., et al. Disability caused by minor head injury. *Neurosurgery* 1981; 9: 221–228.

Sidaros, A., Engberg, A. W., Sidaros, K., et al. Diffusion tensor imaging during recovery
from severe traumatic brain injury and relation to clinical outcome: a longitudinal
study. *Brain* 2008; 131: 559–572.

Smith, D. H., Meaney, D. A., and Shull, W. H. Diffuse axonal injury in head trauma. *J
Head Trauma Rehabil* 2003; 18: 307–316.

Strich, S. J. Diffuse degeneration of the cerebral white matter in severe dementia following head injury. *J Neurol Neurosurg Psychiatry* 1956; 19: 163–185.

Unterharnscheidt, F. About boxing: review of historical and medical aspects. *Texas Rep
Biol Med* 1970; 28: 421–495.

Vannorsdall, T. D., Cascella, N. G., Rao, V., et al. A morphometric analysis of neuroanatomic abnormalities in traumatic brain injury. *J Neuropsychiatry Clin Neurosci* 2010;
22: 173–181.

Wortzel, H. S., Filley, C. M., Anderson, C. A., et al. Forensic applications of cerebral
single photon emission computed tomography in mild traumatic brain injury. *J Am
Acad Psychiatry Law* 2008; 36: 310–322.

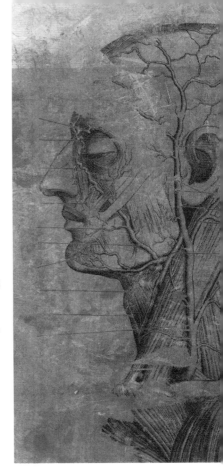

CHAPTER TWELVE

DEMENTIA

The first several chapters of this book were primarily devoted to neurobehavioral syndromes that can be related to focal disruption of cerebral areas concerned with the representation of cognition and emotion. Our review then led to traumatic brain injury (TBI), a problem that, because of its diverse manifestations, produces both focal and diffuse syndromes. To conclude this volume, it is now appropriate to turn to another syndrome, one that by definition represents widespread brain dysfunction: the increasingly threatening problem of dementia. In contrast to the disorders of arousal and attention, dementia does not feature prominent alterations in level of consciousness or attention. In contrast to focal neurobehavioral syndromes, dementia is typically characterized by diffuse or multifocal cerebral damage, and although typical focal syndromes routinely appear as components of dementia, a combination of syndromes adding up to constitute the clinical picture is required for the diagnosis. By far the most common of the persistent neurobehavioral syndromes, dementia assumes special importance because of the prospect of these strongly age-related and mostly irreversible conditions increasing in prevalence as the population of the industrialized world ages. Alzheimer's disease (AD) stands atop the list of dementias that threaten the health of older adults around the globe (Katzman 1976; Cummings 2004).

Perhaps the most frequent patient complaint heard by behavioral neurologists is that of memory loss. Clinicians recognize that any realm of cognitive impairment may account for the problem, but nevertheless the patient usually identifies impaired memory as the reason for consultation, and the physician's task is to assess the nature of the problem and define it more precisely so that diagnostic and therapeutic efforts can proceed. Although many individuals voice this concern, the apprehension about failing memory—real or imagined—is especially evident in the elderly. The well-publicized problem of dementia impels many older adults to worry about lapses in memory, which may represent only a mild decline in memory (Cullum et al. 1990) or sustained attention (Filley and Cullum 1994) that can occur in normal aging. A good general rule is that memory impairment sufficient to interfere with usual social and occupational activities is significant, whereas annoying lapses that can be circumvented by compensatory strategies are not. Clinical evaluation, often with the assistance of neuropsychological testing, can be helpful in this setting to exclude cognitive impairment. If suspicion of incipient dementia persists and reversible causes of cognitive impairment are found and treated, reassurance with follow-up is indicated; dementing illness is usually progressive and will reveal itself in time if it is destined to develop. Depression, of course, must also be considered at this early stage (see below).

For many years, a simple dichotomy between normal aging and dementia was maintained so that older persons with memory complaints could be conveniently classified as belonging to one or the other group. Thus concepts such as *benign senescent forgetfulness* (Kral 1962) and *age-associated memory impairment* (AAMI; Crook et al. 1986) were invoked to formalize the cognitive changes of normal older people in contrast to those characteristic of people with dementia. These categories have some clinical utility in that a normal older person can be reassured by this term, if needed, because it implies that mild memory changes in aging, while real and often frustrating, do not signify a brain disorder. Substantial relief can be provided to older people concerned about failing memory, many of whom will recall the progressive dementia from which a parent suffered, who are found on neurobehavioral evaluation to have normal cognitive function for their age.

As knowledge of this area advanced, however, it became clear that memory and other cognitive changes in the elderly posed a more nuanced clinical challenge. In the late 1990s, the concept of *mild cognitive impairment* (MCI) was introduced to describe older people who exhibit a transitional stage between normal cognitive aging and dementia, specifically AD (Petersen et al. 1999). MCI

TABLE **12.1.** Clinical features of mild cognitive impairment

Memory complaint, preferably corroborated by an informant
Normal activities of daily living
Normal general cognitive function
Abnormal memory for age and education
No dementia

Source: Petersen et al. 1999, 304.

was defined as a condition in which an older person presents with a memory complaint, normal activities of daily living, normal general cognitive function, abnormal memory for age and education, and no dementia (Petersen et al. 1999; Table 12.1). The concept steadily gained widespread, although not universal, acceptance, and MCI is now generally acknowledged to portend an AD risk, for those 65 or older, of 10–15 percent per year, as opposed to a risk of 1–2 percent for those without MCI (Petersen and Negash 2008). Many people with MCI do in fact have early AD pathology, but up to one half of them may have other neuropathologies (Schneider et al. 2009). Further complicating the situation is the recognition that some 20 percent are destined to develop other dementias, and a small number—perhaps 5 percent—revert to normal cognition after some time elapses (Petersen and Negash 2008). These observations have led to a distinction between so-called amnestic MCI, which is more likely to lead to AD, and non-amnestic MCI, which may culminate in a range of other outcomes (Petersen and Negash 2008). This topic remains vigorously studied, and whereas much is still uncertain, MCI has helped establish that a spectrum of cognitive changes can be seen in older people, ranging from normal cognition of aging through MCI and finally to dementia of many types. Effective treatment of MCI remains elusive, however, as the cholinesterase inhibitors—the first agents approved for use in AD—have not been proven effective in warding off the disease (Petersen and Negash 2008), and the MCI concept thus far has its greatest value as identifying a condition that merits enhanced vigilance for the appearance of dementia.

Dementia most simply means an impairment of mental ability (from the Latin *de + mens*) caused by brain dysfunction, implying that a decline in intellect has occurred from a previously stable level. More specific medical definitions, however, have been advanced. The influential psychiatric *Diagnostic and Statistical Manual of Mental Disorders*, now in its fourth edition (DSM-IV), requires the development of multiple cognitive deficits that is sufficient to interfere with normal social or occupational functioning (American Psychiatric Association 1994). The DSM-IV criteria have been influenced by behavioral neurologists who have emphasized identification of key areas of impairment and, by implication, the neuroanatomic basis of these deficits. However, the DSM-IV requires that memory impairment be included as one of the cognitive deficits that characterize the disorder, and this definition may therefore fail to include patients with diseases such as frontotemporal dementia (see below) who do not have memory loss as a presenting feature of the disease. *Dementia* has thus been more usefully defined as an acquired and persistent impairment in intellectual ability that affects at least three of the following nine domains: memory, language, visuospatial perception, praxis, calculations, conceptual or semantic knowledge, executive function, personality or social behavior, and emotional awareness or expression (Mendez and Cummings 2003). It will be recalled that these realms can all be assessed in the mental status examination (Chapter 2), and, like the other syndromes in this book, deficits can be elicited by comprehensive neurobehavioral evaluation.

Dementia has traditionally been assumed to result from diffuse damage to the cerebral cortex, and disruption of "higher cortical function" has often been assumed to be the sole explanation for the cognitive and emotional manifestations of the syndrome. Whereas it is doubtless true that many of the neural operations that account for human behavior take place in the neocortex and hippocampus, it is equally apparent that subcortical regions such as the thalamus, basal ganglia, and cerebral white matter participate in these activities as well. This chapter will review dementia as a syndrome of widespread brain involvement that includes but is not confined to cortical gray matter. Since the brain acts as an integrated whole to produce the entire range of human behavior, dementia can result from afflictions of either cortical or subcortical structures or from a combination of many regional neuropathologies. These disorders will highlight a theme frequently elaborated heretofore regarding the importance of distributed neural networks throughout the brain that mediate specific functions. Dementia illustrates how disruption of these networks at many points can cause identifiable syndromes that coalesce into one multifaceted clinical entity.

The list of conditions that result in dementia is a long one, and new entities are frequently added as they are discovered. In keeping with the emphasis of this book, a classification of the dementias based on the primary regions of neuropathologic involvement is presented in Table 12.2. Steadily accumulating information has underlined the importance of subcortical gray matter diseases in the causation of dementia (Cummings and Benson 1984; Bonelli and Cummings 2008), and cerebral white matter has also been recognized as being commonly associated with the syndrome (Filley et al. 1988; Filley 1998, 2001). Still another group of disorders leads to dementia because of diverse neuropathologic lesions, constituting the category of mixed dementia. This classification will form the basis for considering brain-behavior relationships in dementia; other clinical and basic aspects of dementia are reviewed in detail elsewhere (Mendez and Cummings 2003; Ropper and Samuels 2009).

Whereas this classification has the most relevance for our purposes, clinicians must also recall that dementias can be divided into reversible and irreversible categories. Each of the entities below can also be classified in this fashion,

TABLE 12.2. Major causes of dementia classified by their neuropathological basis

Cortical	Subcortical
Alzheimer's disease	Huntington's disease
Frontotemporal lobar degeneration	Parkinson's disease
Frontotemporal dementia	Progressive supranuclear palsy
Progressive nonfluent aphasia	Wilson's disease
Semantic dementia	Chronic toxic and metabolic disorders
	Depression

White Matter	Mixed
Metachromatic leukodystrophy	Multi-infarct dementia
Multiple sclerosis	Creutzfeldt-Jakob disease
HIV-associated dementia	Dementia with Lewy bodies
Systemic lupus erythematosus	Corticobasal degeneration
Toluene leukoencephalopathy	Neurosyphilis
Cobalamin deficiency	Subdural hematoma
Binswanger's disease	Meningioma
Traumatic brain injury	Hypoxia-ischemia
Glioma	
Normal pressure hydrocephalus	

and of paramount importance is the detection of the reversible dementias as early as possible so that the outcome of treatment can be most favorable. The evaluation and treatment of dementia are also well described in other texts (Mendez and Cummings 2003; Ropper and Samuels 2009).

The classification given in Table 12.2 is useful but not without controversy. The idea of subcortical dementia (see below), presented as a conceptual framework to characterize the dementing illness seen in progressive supranuclear palsy (Albert, Feldman, and Willis 1974) and Huntington's disease (McHugh and Folstein 1975) more than thirty years ago, has met with criticism because of difficulties securely delineating "subcortical" features of dementia, and the fact that the pathology of cortical and subcortical diseases may overlap to a considerable degree (Whitehouse 1986). Nevertheless, dementias do affect the brain in distinct ways that reflect primary sites of neuropathology, and clinical experience teaches that viewing dementia as solely due to cortical degeneration is too simplistic. As expected in brain-behavior relationships, the localization of neuropathology matters more for clinical symptoms and signs than does its specific type, and different areas of involvement produce distinct dementia syndromes (Cummings and Mega 2003). This chapter will illustrate the profiles of impairment that result from cortical, subcortical, white matter, and mixed dementias, using specific diseases, injuries, and intoxications as examples. The dementias are providing ample demonstration that the term "higher cortical function" is too limiting and that "higher cerebral function" is clearly preferable.

CORTICAL DEMENTIAS

The most important dementia on any list, cortical or otherwise, is *Alzheimer's disease* (AD). First described more than a century ago by Alois Alzheimer in a fifty-five-year-old woman who had a progressive four-year course of amnesia, aphasia, and personality change (Alzheimer 1907), the disease has been known since then as a cortical degenerative process featuring neuritic plaques and neurofibrillary tangles in addition to widespread neuronal and synaptic loss (Cummings 2004; Querfurth and LaFerla 2010). Since Alzheimer's description, AD has evolved from an obscure clinicopathologic entity to one of the major medical problems of the industrialized world. By far the most common of the progressive dementias, accounting for more than 50 percent of all dementia cases in older people, AD affects millions around the world, entails enormous medical costs, and is the primary reason for elderly people to be admitted to nursing homes (Katzman 1986; Cummings 2004).

AD is a progressive neurodegenerative disease of unknown etiology except in a small number of cases that begin before age sixty-five and are inherited as an autosomal dominant trait (Cummings 2004). Alzheimer's first patient was likely a familial case, but today the vast majority of patients are older than sixty-five and have sporadic AD, which is influenced, but not caused, by a susceptibility gene known as apolipoprotein ε4 (APOE4; Cummings 2004). Alzheimer reported his case as "presenile dementia," but as the neuropathology of AD is now recognized to be identical at all ages, the distinction between "presenile" and "senile" forms is unjustified (Katzman 1986; Cummings 2004). The disease is still incurable, and from the time of onset, its duration until death approximates 6–12 years; no cases of remission have been reported (Mendez and Cummings 2003).

The disease affects the hippocampus and adjacent areas early in the course, as shown in Figure 12.1, accounting for the amnesia that nearly always heralds the onset. The parietal and temporal cortices (Figure 12.1) are also selectively damaged at an early stage, leading to deficits in language, praxis, calculations, perception, and visuospatial skills (Mendez and Cummings 2003). Thus the classic cortical syndromes of aphasia (transcortical sensory or Wernicke's), apraxia (ideomotor or ideational), and agnosia (object agnosia or prosopagnosia) soon accompany the initial amnesia (Mendez and Cummings 2003). The frontal association cortices are affected somewhat later, but primary sensory and motor cor-

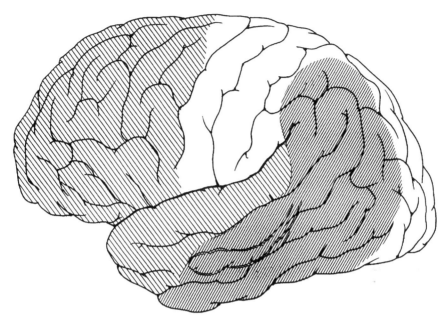

FIGURE 12.1. *Cerebral areas selectively affected in early Alzheimer's disease (dark shading: severe involvement; light shading: moderate involvement).*

tices are preserved until very late in the disease. It is the selective attack on hippocampal and association cortices that accounts for the specific neurobehavioral presentation of this disease.

As AD advances, a fairly predictable sequence of progressive clinical decline can be observed (Mendez and Cummings 2003). Amnesia remains a prominent feature, and its worsening leads to greater dependence on caregivers. Other cognitive deficits also become more evident, and problems with driving, financial competence, and home safety may appear. Personality change and paranoid delusions may appear transiently. Sphincteric incontinence eventually develops, and at this stage many patients require nursing-home care. Finally, after many years of the illness, patients become bedridden, mute, and totally dependent; the minimally conscious state (MCS; Giacino et al. 2002) and later persistent vegetative state (PVS; Multi-Society Task Force on PVS 1994) can ensue if no other cause of mortality intervenes. Death in AD usually occurs as a result of a pulmonary, urinary tract, or decubitus ulcer infection.

The cause of AD remains the central unanswered question for most patients. Families with early-onset autosomal dominant AD were first linked with Down syndrome (trisomy 21) by the universal development of AD neuropathology in Down syndrome patients who reach midlife, and research led to the identification of the amyloid precursor protein (APP) gene on chromosome 21 (St. George-Hyslop et al. 1987). Other investigators found that mutations in presenilin-1 on chromosome 14 (Schellenberg et al. 1992) and presenilin-2 on chromosome 19 (Corder et al. 1993) can also produce early-onset autosomal dominant AD. Testing for mutations in all these genes is now available (Howard and Filley 2009). The great majority of AD patients, however, have sporadic disease, and the etiology is unknown. Intoxication with aluminum (Perl and Brody 1980) and prion infection (Prusiner 1993) have been suggested but not confirmed. A plausible view is that the cause of sporadic AD includes a genetic contribution, probably involving a variety of nuclear genes (Clark and Goate 1993; Querfurth and LaFerla 2010) or even the mitochondrial genome (Parker et al. 1994; Swerdlow and Khan 2009), together with environmental factors such as TBI and low educational attainment (Cummings 2004). Most recently, recognition of the combination of AD and vascular pathology in many demented patients has raised the idea that vascular mechanisms may contribute to AD (Querfurth and LaFerla 2010); one proposal has been advanced that primary white matter pathology—from ischemia, TBI, and other factors—initiates the pathogenesis of AD (Bartzokis 2009). These ideas indicate that efforts to prevent or treat hypertension, diabetes mellitus, heart disease, obesity, smoking, hypercholesterolemia, and TBI may add additional benefit in mitigating the scourge of AD.

Efforts to treat AD pharmacologically have centered mostly on the well-accepted cholinergic cell loss in a region within the basal forebrain known as the nucleus basalis of Meynert (Whitehouse et al. 1982). Effective therapy has been discouragingly elusive, but the cholinesterase inhibitors and the neuroprotective drug memantine have offered some possibility of symptomatic treatment. Tacrine was the first approved cholinesterase inhibitor and was followed by donepezil, rivastigmine, and galantamine (Cummings 2004). Memantine was soon added to the AD pharmacopaeia as an agent designed to reduce glutamate-induced neuronal excitoxicity (Cummings 2004). Today one of the cholinesterase inhibitors and memantine are typically prescribed together in an effort to slow the pace of cognitive decline, but it must be recognized that these are not disease-modifying drugs and the disease is still irreversible. Irreversibility, however, does not imply untreatability, and much can be done for patients and their families with informed and sympathetic counseling that can significantly lighten the burden of this dreaded illness.

Curative medical therapy of AD must await a better understanding of its etiology. Ever since Alzheimer's report (1907), attention has naturally been drawn to the neuropathological hallmarks of the disease: neuritic plaques and neurofibrillary tangles (Figure 12.2). Recently, a major effort has centered on neuritic

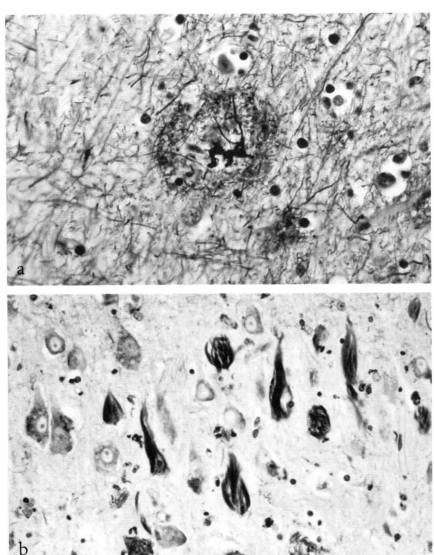

FIGURE 12.2. *Microscopic appearance of the neocortex from an AD brain showing (a) a neuritic plaque and (b) neurofibrillary tangles.*

plaques in view of much evidence that their major constituent, a protein called amyloid β42, particularly in oligomeric form, initiates a cascade of neurotoxicity and causes AD (Cummings 2004). However, several problems have hampered progress. Although it was long thought that the major correlate of intellectual loss in AD was the presence of neuritic plaques (Blessed, Tomlinson, and Roth 1968), cortical synapse loss is now accepted as the strongest structural correlate of dementia severity (Terry et al. 1991). Moreover, it is evident that many older people harbor significant amounts of cerebral amyloid and remain cognitively normal (Alzenstein et al. 2008). Last, and most important, efforts to treat AD with immunologic agents that target amyloid have so far failed to improve cognition even as they clear amyloid from the brain (Querfurth and LaFerla 2010). Thus the dominant theory of AD pathogenesis—the amyloid cascade hypothesis—has significant inconsistencies that require further investigation (Hardy 2009).

A reasonable current formulation of the etiology of AD is that multiple factors, genetic and environmental, contribute to the loss of synapses that is the final common cause of clinical symptoms and symptoms. If this is true, then intellectual capacity depends critically on the number of cortical synapses, and efforts to maintain and even increase these numbers are reasonable. In this regard, there is intriguing evidence that higher educational and occupational level may be a protective factor against the development of AD, implying that exposure to intellectual challenges—or "mental exercise"—can increase synaptic density and forestall the appearance of dementia (Stern 2009). Animal studies showing that complex environments favor dendritic growth and increased brain weight (Marx 2005; Stern 2009) add further support to this idea. In contrast, once the disease is established, the decline in function may be unaffected by premorbid educational achievement (Filley and Cullum 1997). Thus there is growing interest in the possibility that both physical and mental exercise can help prevent AD, particularly if pursued early in life and continued regularly (Marx 2005). At least until a cure becomes available, the advice to maintain optimal physical and mental health on a consistent basis seems eminently reasonable. From our perspective, these considerations illustrate that microscopic intracortical events underlie the cognitive changes in AD, and a behavioral neurology at the neuronal and synaptic levels in AD is emerging that promises to reveal insights into brain-behavior relationships as surely as the classic work of Broca, Wernicke, and others of the nineteenth century who explored the effects of focal cerebral lesions.

Frontotemporal lobar degeneration (FTLD) is the other and less common entity in the cortical dementia category (Josephs 2007). This disorder has a complex history, as until recently it was known as *frontotemporal dementia* (FTD), the modern term for what was formerly known as Pick's disease (Pick 1892). FTD persists as a common diagnosis today and is now seen as one of three forms of FTLD, the others being *progressive nonfluent aphasia* (PNFA) and *semantic dementia* (SD). The term *primary progressive aphasia* (PPA) is sometimes used to include both PNFA and SD (Mesulam 2007). In general, FTLD shows a different pattern of cortical degeneration that produces correspondingly distinct neurobehavioral profiles

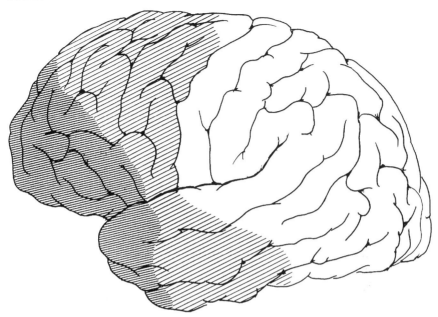

FIGURE 12.3. *Cerebral areas selectively affected in early FTD (dark shading: severe involvement; light shading: moderate involvement).*

reflecting frontal and temporal involvement. In contrast to AD, the hippocampus is spared initially in FTLD, whereas the frontal and/or anterior temporal cortices are heavily damaged. This lobar pattern is particularly evident in FTD, the most disabling of the three FTLD variants, in which bilateral frontotemporal degeneration is evident (Figure 12.3). This disorder is also referred to as *behavioral variant FTD* to emphasize the prominence of its behavioral manifestations. Although the damage is bilateral, recent evidence suggests that right frontal degeneration is most responsible for behavioral dysfunction (Rosen et al. 2005; Sollberger et al. 2009), and the existence of a right frontal network for moral behavior has been proposed (Mendez 2009). In PPA and SD, more restricted degeneration occurs, in the left frontal and temporal lobes, accounting for specific language impairments. In PPA, there is early anomia advancing to nonfluent aphasia, and in SD, fluent aphasia develops with a characteristic loss of what words mean. In all cases, the degeneration eventually advances to produce more severe clinical and neuropathological involvement.

The understanding of FTLD genetics and molecular pathology has advanced rapidly in recent years. It is now known that up to 50 percent of FTLD cases are familial and often seen in association with other neurodegenerative diseases, including amyothrophic lateral sclerosis (Bigio 2008). The histopathology of FTLD can represent one of three forms of proteinopathy: (1) FTLD-tau, the most common, is a tauopathy associated with microtubule-associated protein tau mutations on chromosome 17 and includes classic Pick's disease; (2) FTLD-TDP features TAR DNA-binding protein-43 and is associated with progranulin gene mutations on chromosome 17; and (3) FTLD-FUS (for "fused in sarcoma") is associated with FUS protein and mutations on chromosome 16 (Cairns and Ghoshal 2010). These distinctions highlight the emerging prominence of genetics in FTLD and suggest that further efforts to investigate genetic etiology in neurodegenerative diseases will prove rewarding.

Genetics and histopathology, of course, are typically unknown to clinicians seeing affected patients, but the regional predilection of FTLD for frontal and/or temporal regions usually allows the clinical distinction of all the variants of this disease from AD. A key clinical point is that manifestations of frontal and temporal lobe dysfunction predominate early in the course while memory is still unaffected (Filley and Cullum 1993; Filley, Kleinschmidt-DeMasters, and Gross 1994; Mendez and Cummings 2003). In contrast to the early amnesia, fluent aphasia, and relatively preserved personality of patients with AD, in FTLD there is early disinhibition or nonfluent aphasia without initial memory loss. Here is a noteworthy example of a dementing illness in which memory dysfunction is *not* a presenting problem, and it illustrates the point that dementias each have a clinical "signature" when subjected to careful analysis. Diagnosis can be challenging, but detailed neurobehavioral evaluation in concert with neuropsychological testing and neuroimaging techniques can be revealing, particularly early in the disease course (Filley and Cullum 1993). Later, when the degeneration is more advanced, the neurobehavioral distinction between AD and FTLD is more difficult. As in AD, no cure is available, but in contrast to the more common cortical dementia, no pharmacotherapy has been approved for clinical use in FTLD.

SUBCORTICAL DEMENTIAS

The entity of *subcortical dementia* has been an organizing concept for behavioral neurology for many decades (Cummings and Benson 1984), but controversy persists about its usefulness. More recent formulations of the idea have altered the term to *frontal-subcortical dementia*, recognizing that frontal systems are often disrupted by the primary subcortical gray matter pathology (Bonelli and Cummings 2008). The discussion of frontal-subcortical circuits in Chapter 1 sheds some light on how frontal systems may be involved in the pathogenesis of subcortical dementias by virtue of the extensive connectivity within these areas. Whatever terminology is adopted, the idea offers considerable utility as a clinical and neuroanatomic concept, and many clinicians point out that these dementias have more in common with each other than they do with AD (Bonelli and Cummings 2008). Originally, the subcortical dementias included diseases of both the subcortical gray matter, such as Parkinson's disease, and the white matter, such as multiple sclerosis (MS), and this general classification still has merit (Bonelli and Cummings 2008). However, in view of the emerging category of white matter disorders and their effects on higher function, a case can be made—especially in a book about the neuroanatomy of behavior—for separating the white matter disorders from those that affect the subcortical gray matter. Our approach will be to consider both the subcortical dementias—those specifically affecting only the subcortical gray matter—and the white matter dementias that arise from damage to myelinated systems. Evidence justifying this distinction on clinical grounds will also be presented. Further study will undoubtedly refine these distinctions, but there is ample reason to proceed from the notion that dementias present differing patterns of behavioral alteration because of the varying distribution of affected cortical and subcortical regions.

When the topic became current four decades ago, the first diseases listed as subcortical dementias were *Parkinson's disease* (PD), *Huntington's disease* (HD), *progressive supranuclear palsy* (PSP), and *Wilson's disease* (WD). Standard textbooks address the neurology of these conditions (Ropper and Samuels 2009), but, in brief, they all share dementia in association with some variety of movement disorder, and all are due mainly to pathology in the basal ganglia and related subcortical gray matter structures (Figure 12.4).

The role of the subcortical gray matter in neurobehavioral function intrigued neurologists for many decades. Although these structures are indeed gray matter regions, they have been more clearly linked with motor function than with cognition or emotion. PD, for example, is a common and disabling movement disorder characterized by resting tremor, bradykinesia, and rigidity, and these motor disturbances are strongly associated with loss of pigmented dopaminergic neurons in the substantia nigra (Figure 12.5).

Similarly, HD, the autosomal dominant disease featuring chorea and dementia, regularly shows atrophy of the caudate nuclei (Figure 12.6). In both PD and HD, the degenerative process proceeds over time ineluctably to involve other regions, including the cortex, but the earliest and so far most readily treated clinical features stem from the neuropathology in selected subcortical gray matter regions.

Yet despite the motor abnormalities being readily recognized and remaining the clinical hallmarks of these diseases, dementia has been appreciated as a cause of significant morbidity in these and other similar disorders. As will be discussed later in this chapter, the work derived from the observations supporting subcortical dementia has served to facilitate much progress in the understanding of the neural basis of all the dementia syndromes.

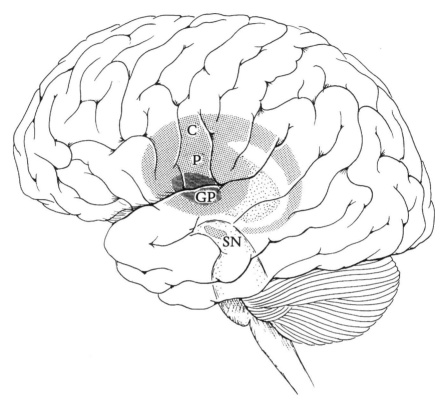

FIGURE **12.4.** *Basal ganglia primarily affected in subcortical dementias (C: caudate; P: putamen; GP: globus pallidus; SN: substantia nigra).*

On somewhat less secure grounds, *chronic toxic and metabolic disorders* cause neurobehavioral impairment that can be viewed as subcortical dementia (Mendez and Cummings 2003). This notion suggests that the acute confusional state (Chapter 3) can become chronic if the insult is persistent, and that this "chronic confusional state" has clinical and pathological similarities to subcortical dementia. Finally, the *dementia syndrome of depression*—contentious both because many prefer to separate depression from dementia altogether and because its neuropathologic anatomy is still largely mysterious—has been included in this group (King and Caine 1990).

Clinical research on the original four entries has been actively pursued. The first modern formulation of subcortical dementia described four cardinal features of patients with PSP: slowness of thought processes, forgetfulness, personality changes including apathy and depression, and impaired ability to manipulate acquired knowledge (Albert, Feldman, and Willis 1974). A similar

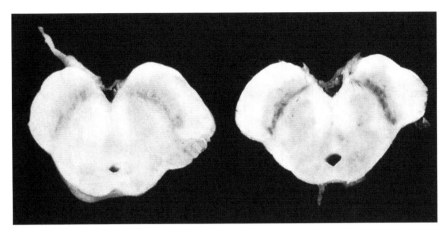

FIGURE 12.5. *Section of the midbrain of a PD patient showing loss of pigmented cells in the substantia nigra (A) compared with a normal midbrain of the same age (B).*

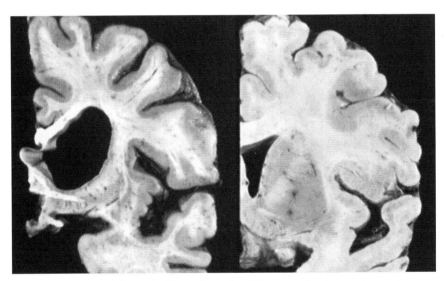

FIGURE 12.6. *Section of the cerebral hemisphere of an HD patient showing massive caudate atrophy (A) compared with a normal brain of the same age (B).*

clinical profile was noted a year later in patients with HD (McHugh and Folstein 1975). Neuropsychological studies of patients with PD (Cummings 1988) and WD (Medalia, Isaacs-Glaberman, and Sheinberg 1988) also found deficits of this nature. These impairments, primarily indicating problems in the timing and activation of cortical processes rather than disruption of the cortex itself, have been interpreted as representing *fundamental* deficits in the areas of arousal, attention, mood, and motivation (Albert 1978). In contrast, the *instrumental* functions of memory, language, praxis, and perception have been regarded as intact in subcortical dementias, but not in cortical dementias, where amnesia, aphasia, apraxia, and agnosia are common (Albert 1978). Despite the seeming vagueness of the clinical features of subcortical dementia and the presence of neuropathology that can extend, as in PD, into the cerebral cortex, several studies have documented measurable neuropsychological differences between various cortical and subcortical dementias (Huber et al. 1986; Pillon et al. 1986; Brandt, Folstein, and Folstein 1988). Precise distinction of these syndromes can still be problematic for the clinician, but the concept of relatively specific neurobehavioral deficits generated by subcortical pathology has gained credibility (Huber and Shuttleworth 1990; Drebing et al. 1994; Bonelli and Cummings 2008). It must be pointed out, however, that dementing diseases frequently overlap (Mendez and Cummings 2003; Querfurth and LaFerla 2010), so that a case of "pure" subcortical dementia may be more the exception than the rule. PD is often combined with AD at autopsy, for example, and it is not surprising that neurobehavioral features of dementia during life may be difficult to assign to the subcortical or cortical dementia category.

Most controversial is the dementia syndrome of depression, although there is no doubt that depression is a common disorder that often enters into the differential diagnosis of amnesia or dementia. The term "pseudodementia" has been applied to this condition (Wells 1979), but suffers from the drawback that the dementia syndrome of depression is in fact a true dementia—with all the functional disability implied by the term. Dementia as defined above (Mendez and Cummings 2003) can occur in severe depression (Caine 1981), and the common assumption that depression is a "functional" illness as opposed to the "organic" syndrome of dementia is outmoded and of no clinical utility. Undeniably, the multiple cognitive deficits of dementia can develop in the context of depression, and these may relate to structural or neurochemical pathology in ascending subcortical systems concerned with the regulation of mood (King and Caine 1990). Patients with depression are often slow, inattentive, forgetful, and unmotivated, and their clinical history and predictably poor performance on mental status examination justify the term dementia as surely as in a patient with AD. Dementia, as emphasized above, does not imply irreversibility, and the recognition of the reversible dementia syndrome of depression is one of the most important tasks the behavioral neurologist is called upon to perform.

Before leaving the topic of depression, a brief review of the cerebral basis of *mood disorders* is in order. A summary statement can be tentatively put forth that subcortical structures together with their frontal and temporal lobe connections are implicated. Because of the generally efficacious pharmacologic treatment of depression and bipolar disorder, theories of pathogenesis have centered on disturbances of neurotransmitter systems rather than studies on structural changes in the brain. Early investigations of the brain in mood disorders disclosed nonspecific findings—ventriculomegaly, sulcal widening, and hypofrontality on functional imaging studies (Jeste, Lohr, and Goodwin 1988). Studies of secondary mood disorders—those due to neurologic disorders such as TBI, stroke, brain tumor, epilepsy, MS, and degenerative disease—further implicated the frontal lobes, temporal lobes, and basal ganglia (Cummings 1993; Guze and Gitlin 1994). As for the subtypes of mood disorder, depression has been associated with frontal lobe dysfunction from functional neuroimaging studies (Koenigs and Grafman 2009). Meanwhile, bipolar disorder has been linked with frontal and subcortical hypometabolism (Bearden, Hoffman, and Cannon 2001), and white matter abnormalities have also been revealed (Osuji and Cullum 2005). Finally, an interesting body of evidence has suggested that mania is associated with dysfunction of anterior right hemisphere structures (Cummings 1986; Starkstein et al. 1990; Paskavitz et al. 1995).

From a clinical point of view, a useful broad generalization is that psychiatric disorders are more likely to occur with dysfunction of subcortical than of cortical structures (Salloway and Cummings 1994). The fundamental functions of arousal, attention, and mood are organized primarily by brainstem, diencephalic, and limbic regions, and damage to these subcortical areas is associated with a wide variety of behavioral alterations such as depression and psychosis that are traditionally viewed as psychiatric in nature. In contrast, the cortex can be regarded as performing more specific cognitive operations such as memory, language, praxis, perception, and executive function. Such a view implies that the understanding of psychiatric illness in general may be found in the study of subcortical systems, and perhaps the term *neuropsychiatric* is a better descriptor of the effects of subcortical disease (Salloway and Cummings 1994). However, it is clear that the cortex and subcortex are heavily interconnected, and as a result both clinical and neuroanatomic distinctions are only approximate. In keeping with emerging understanding of distributed neural networks and frontal-subcortical circuits, our last word must be that the functions of the cortical and subcortical regions must be seen as complementary and not dichotomous.

WHITE MATTER DEMENTIAS

The cerebral white matter occupies approximately one half the volume of the adult cerebrum (Miller, Alston, and Corsellis 1980) and some 135,000 kilometers of myelinated fibers course within and between the hemispheres (Saver 2006). White matter functions to connect cortical and subcortical areas within and between the hemispheres, facilitating rapid and efficient interregional communication by the substantial increase in conduction velocity conferred by the myelin

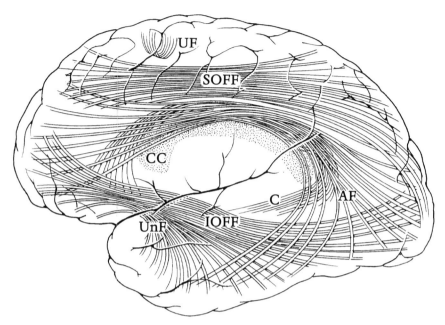

FIGURE 12.7. *Cerebral white matter tracts primarily affected in white matter dementias (UF: U fibers; CC: corpus callosum; SOFF: superior occipitofrontal fasciculus; IOFF: inferior occipitofrontal fasciculus; AF: arcuate fasciculus; UnF: uncinate fasciculus; C: cingulum).*

investing most cerebral axons. Although behavioral neurologists have theorized about white matter tracts and their significance (Wernicke 1874; Geschwind 1965), this major neuroanatomic component of the brain has been relatively neglected in comparison with the scrutiny given to cortical and subcortical gray matter. With the advent of MRI, a technique providing detailed differentiation of gray and white matter areas, elegant in vivo imaging is helping to expand our knowledge of brain diseases that alter behavior by virtue of their assault on brain white matter. In parallel, a greater understanding of the role of white matter in higher function is also developing. It has recently been observed, for example, that the frontal lobe white matter volume is disproportionately larger in humans than in other primates, implying that white matter has a played a special role in human evolution (Schoenemann, Sheehan, and Glotzer 2005).

There are three major constituents of the cerebral white matter: projection fibers, commissural fibers, and association fibers (Nolte 2002). Projection fibers are involved with elemental motor and sensory systems, and the fibers most important for our purposes are the commissural and association fibers, depicted in Figure 12.7. The major commissural tract, of course, is the corpus callosum, a massive interhemispheric structure connecting all cerebral areas, but connections between the hemispheres are also made by the anterior and hippocampal commissures. The intrahemispheric association fibers are more complex. First, there are short association fibers, also known as arcuate or U fibers, which connect adjacent cortical gyri. In addition, there are five long association fiber bundles, each of which connects the frontal lobe with one or more of the other cerebral lobes. These are the arcuate (superior longitudinal) fasciculus, the superior occipitofrontal fasciculus, the inferior occipitofrontal fasciculus, the uncinate fasciculus, and the cingulum. While these standard descriptions of white matter anatomy are still useful, the neuroanatomy of white matter is currently undergoing a substantial refinement as new data are collected (Schmahmann and Pandya 2006). With the appearance of new data and the advent of ever more revealing neuroimaging techniques, the precise identification of white matter tracts and their correlations with cognitive and emotional functions will be increasingly clarified.

The importance of white matter in elemental motor and sensory capacities is indisputable, but it has been less clear how or even if white matter contributes to higher function. While important synaptic events such as neurotransmitter release (Chapter 1) and long-term potentiation (LTP; Chapter 4) clearly occur in the hippocampus and neocortex, clinical evidence as well as theoretical considerations compel the view that behavioral neurology risks a perilous omission if the white matter is disregarded. As a general statement, the white matter—by virtue of the *macroconnectivity* established by tracts coursing within and between the hemispheres—can be considered responsible for *information transfer*, while gray matter subserves *information processing* by means of the *microconnectivity* at billions of synapses; both are necessary for normal brain function (Filley 2005).

One familiar research area in which white matter is directly involved, for example, is the investigation of hemispheric disconnection by sectioning of the corpus callosum, usually performed for the relief of intractable epilepsy. Several decades of study have now demonstrated that interhemispheric cognitive integration is indeed mediated by the corpus callosum (Seymour, Reuter-Lorenz, and Gazzaniga 1994). This kind of surgery, however, and clinical instances of callosal syndromes such as pure alexia (Chapter 5) and callosal apraxia (Chapter 6), are uncommon. In contrast, primary diffuse involvement of the cerebral white matter does occur in a substantial number of diseases and with TBI. Without exception, these disorders have all been reported to result in some form of neurobehavioral impairment (Filley 1998, 2001). Dementia is the most common of the cerebral white matter neurobehavioral syndromes and will be reviewed in some detail.

Ten categories can be conveniently established in considering the white matter disorders that may result in cognitive impairment or dementia: genetic, demyelinative, infectious, inflammatory, toxic, metabolic, vascular, traumatic, neoplastic, and hydrocephalic (Filley 2001). Examples of each of these will be discussed below (Table 12.2), but it is worth noting that well over 100 diseases, injuries, or intoxications are recognized that have prominent or exclusive effects on cerebral white matter (Filley 2001). Moreover, careful study of the clinical manifestations of these diseases discloses that, in addition to the elemental neurologic deficits that are often prominent, neurobehavioral dysfunction has, without exception, been observed to be a potential sequela of these disorders. Dementia or cognitive impairment, while by no means invariable, may certainly occur, and the concept of white matter dementia merits consideration as a relatively unrecognized area of behavioral neurology (Filley et al. 1988; Filley 1998, 2001). We shall consider examples of each category of white matter disorder to illustrate the importance of white matter in neurobehavioral dysfunction across a wide range of neuropathology.

The genetic disease of myelin known as *metachromatic leukodystrophy* (MLD) can rarely present in adolescence or adulthood, and dementia has been described (Shapiro et al. 1994). MLD is an autosomal recessive disease of myelin formation due to a deficiency of the enzyme aryl sulfatase A (Austin et al. 1968). Magnetic resonance imaging reveals extensive cerebral dysmyelination in MLD in advanced cases (Filley and Gross 1992). In addition to dementia, psychosis is strongly associated with late-onset MLD, perhaps because of involvement of frontal and limbic system white matter (Filley and Gross 1992; Hyde, Ziegler, and Weinberger 1992).

Multiple sclerosis (MS) is a demyelinative disease of the central nervous system well recognized by neurologists as a cause of neurological disability. Although descriptions of cognitive and emotional dysfunction in MS were published over a century ago by Jean Martin Charcot (1877), the many neurobehavioral manifestations of the disease received scant attention for much of the twentieth century. The availability of MRI helped change this situation (Goodkin, Rudick, and Ross 1994), permitting correlations between neurobehavioral syndromes and white matter lesions. Cognitive impairment or frank dementia has been most often described, and a variety of neuropsychiatric and focal demyelination syndromes may also be encountered. Community-based neuropsychological studies of MS found a prevalence of cognitive impairment of 43 percent (Rao et al. 1991), and dementia has been described in 23 percent (Boerner and Kapfhammer 1999). Cognitive deficits are especially common in more severe forms of the disease (Heaton et al. 1985). These deficits may be overlooked with the use of the Mini-Mental State Examination (MMSE; Folstein, Folstein, and McHugh 1975), as this test is insensitive to dementia in MS (Franklin et al. 1988). MRI studies of MS disclose a relationship between the severity of dementia and white matter disease burden (Filley 2001), and the dementia of MS differs from that of AD by manifesting as attentional and executive dysfunction rather than amnesia and aphasia

(Filley et al. 1989). Although white matter tract involvement is clearly implicated in cognitive impairment, recent evidence has also documented demyelination in the cerebral cortex (Stadelmann et al. 2008). The role of cortical demyelination is undergoing further study with regard to the pathogenesis of neurobehavioral involvement in patients with MS.

The acquired immunodeficiency syndrome (AIDS) has been recognized to be frequently associated with dementia, a syndrome first named the *AIDS dementia complex* (ADC; Navia et al. 1986a) and later known as *human immunodeficiency virus (HIV)–associated dementia* (HAD; Boissé, Gill, and Power 2008). A striking syndrome of cognitive slowing, inattention, memory loss, and apathy was evident early in the AIDS epidemic, and although many opportunistic infections and neoplasms can involve the brain in AIDS patients to produce a similar picture, dementia was also discovered to result from HIV in the absence of other complications (Navia et al. 1986). The pathogenesis of HAD is complex, as HIV does not directly infect neurons (Boissé, Gill, and Power 2008), and the neuropathologic basis of dementia is not well explained by loss of cortical neurons (Seilhean et al. 1993). Involvement of the basal ganglia is likely, and the subcortical white matter is particularly involved early and throughout the disease (Navia, Jordan, and Price 1986b; Price et al. 1988; Budka 1991). The role of white matter neuropathology in HAD cannot be dismissed in light of evidence that cognitive improvement can be seen in parallel with reduction of MRI white matter disease burden as the infection is treated (Thurnher et al. 2000)

The inflammatory autoimmune diseases of the brain are similar to most infectious diseases in that their pathology is not confined only to the cerebral white matter. The neuropathology of these diseases involves many brain regions as well as many different pathogenic processes. Nevertheless, growing evidence implicates a role for white matter involvement in neurobehavioral dysfunction. *Systemic lupus erythematosus* (SLE) is the best-studied example of this category. Neuropsychiatric lupus refers to a group of syndromes in SLE patients that includes cognitive dysfunction (West 1994), a problem that can be noted in SLE patients both with and without overt neurologic disease such as a stroke (Kozora et al. 1996). White matter hyperintensities on MRI are common in SLE (Kozora et al. 1996), presumably related to the underlying neuropathology of the disease, and a relationship between dementia and leukoencephalopathy in SLE has been suggested (Kirk, Kertesz, and Polk 1991). Data from studies with the newer technique of magnetic resonance spectroscopy (MRS) have shown that even in SLE patients with normal white matter on conventional MRI and no neuropsychiatric features, subtle cognitive impairment correlates with increased choline in the white matter, but not with the neuronal marker N-acetyl aspartate (NAA) or cerebral atrophy (Filley et al. 2009). Thus, support for a contribution of cerebral myelin damage to cognitive impairment in SLE is accumulating. White matter damage may be the earliest effect of SLE in the brain, and involvement of gray matter regions including the hippocampus and neocortex may occur later in the disease.

A particularly convincing example of white matter dementia is in the toxic category: the syndrome of *toluene leukoencephalopathy* consequent to chronic toluene abuse. This unfortunate recreational practice has been securely identified as a cause of dementia by virtue of widespread injury to cerebral white matter (Rosenberg et al. 1988; Filley, Heaton, and Rosenberg 1990). In addition, a variety of corticospinal, cerebellar, and ocular signs have been documented (Hormes, Filley, and Rosenberg 1986). Toluene, an inexpensive and readily available euphoriant found in spray paints and solvents, is often inhaled in large quantities for protracted periods that may extend for years. By virtue of its lipophilicity, toluene has a high affinity for the brain (Hormes, Filley, and Rosenberg 1986), and white matter is selectively affected while gray matter in the cortex and the subcortical nuclei is spared (Rosenberg et al. 1988). The dementia syndrome has a typical subcortical pattern, with inattention, apathy, forgetfulness, and preservation of language (Hormes, Filley, and Rosenberg 1986). MRI is sensitive to the white matter changes, showing loss of differentiation between gray and white matter and increased periventricular white matter signal intensity (Filley, Heaton, and Rosenberg 1990). The association of these MRI abnormalities with dementia is bolstered by their significant correlation with the severity of dementia as determined by neuropsychological testing (Filley, Heaton, and Rosenberg 1990). Toluene leukoencephalopathy is a specific example of the larger spectrum of toxic leukoencephalopathy, a disorder of cerebral white matter related to a wide variety of therapeutic agents, drugs of abuse, and environmental toxins (Filley and Kleinschmidt-DeMasters 2003).

Before leaving the toxic leukoencephalopathies, *alcohol abuse* deserves comment. Alcohol is associated with a plethora of effects on the nervous system, many of which are due to thiamine deficiency, coexistent systemic illness, and TBI. The existence of alcoholic dementia has been debated, some maintaining that only amnesia related to the Wernicke-Korsakoff syndrome (Chapter 4) has been clearly documented (Victor, Adams, and Collins 1989), and others contending that dementia unrelated to thiamine deficiency does indeed exist (Lishman 1981; Mendez and Cummings 2003). Consistent with the latter view, substantial evidence implicates selective loss of cerebral white matter both in laboratory animals exposed to alcohol (Hansen et al. 1991) and in humans with alcoholism (de la Monte 1988; Jensen and Pakkenberg 1993). The reversal of cerebral atrophy as measured by computed tomography and MRI scans after alcohol abstinence (Carlen and Wilkinson 1987; Gazdzinski et al. 2010) also argues for a role of white matter damage in producing alcoholic dementia.

Cobalamin (vitamin B$_{12}$) deficiency is a known cause of cerebral white matter damage (Adams and Kubik 1944), in addition to its well-known effects on the spinal cord, optic nerves, and peripheral nervous system. Focal demyelinative lesions in the brain are similar to those in the dorsal and lateral columns of the spinal cord typical of subacute combined degeneration. A common cause of cobalamin deficiency is pernicious anemia with failure of intrinsic factor production. Involvement of the nervous system can occur, however, even before the onset of hematologic abnormalities. Dementia has been described as one of many neurologic and psychiatric manifestations of cobalamin deficiency (Lindenbaum et al. 1988), and prominent features include mental slowing, memory impairment, and depression (Mendez and Cummings 2003). Lesions of the cerebral white matter are likely responsible for the dementia, and treatment with parenteral vitamin B_{12} may be beneficial if the deficiency has not been prolonged. Well-described cases document improvement in both cognitive dysfunction and MRI white matter hyperintensity with B_{12} replacement (Chatterjee et al. 1996; Stojsavljević et al. 1997).

Binswanger's disease (BD) is an old clinical entity (Binswanger 1894) that has more recently come to life again because of the sensitivity of CT and MRI to ischemic white matter changes in the elderly brain (Hachinski, Potter, and Merskey 1987; Filley et al. 1989). BD is traditionally regarded as a dementia in hypertensive patients who have ischemic demyelination of the hemispheres, frequently accompanied by lacunar infarctions (Babikian and Ropper 1987; Caplan 1995). It is likely that these seemingly ubiquitous white matter changes, which have been given the name *leukoaraiosis* (Hachinski, Potter, and Merskey 1987), represent an incipient form of BD (Román 1996). White matter lesions are associated with hypertension and other cerebrovascular risk factors, and the opportunity for prevention of dementia by managing these risks offers much hope. Many studies have indicated that in both demented persons (Almkvist et al. 1992) and normal elderly (Junque et al. 1990; Ylikoski et al. 1993; Schmidt et al. 1993) white matter changes adversely affect attention and speed of information processing, and these deficits may only occur when a threshold area of white matter involvement has been exceeded (Boone et al. 1992). The evidence thus seems to suggest that white matter changes on CT or MRI represent a neuropathologic change that may in many cases lead to BD. Avoiding the controversies of the term BD, many have adopted the newer terminology of subcortical ischemic vascular dementia (Chui 2007) and the vascular cognitive impairment (Selnes and Vinters 2006) that may precede it, but the main point is that cerebrovascular white matter changes can in many cases lead to white matter dementia (Filley 2001). Multi-infarct dementia, a more common form of vascular dementia, is considered in the next section.

TBI has many effects on the brain, as reviewed in Chapter 11, and inclusion of TBI within the category of white matter dementia is not intended to ignore the other neuropathologies that occur. However, the most common and consistent neuropathological finding in TBI is damage to cerebral and upper brainstem white matter in the form of diffuse axonal injury (DAI; Adams et al. 1982; Alexander 1995), alternatively known as traumatic axonal injury. Although DAI clearly involves axonal damage, there is also significant disruption of myelin, and cell bodies in the gray matter are largely spared. Thus the injury of DAI falls most heavily on white matter regions, and this kind of injury likely accounts for the persistent inattention and difficult emotional alterations that characterize post-traumatic dementia (Filley 2001). Moreover, it is likely that DAI is the most important cause of an entire spectrum of post-traumatic neurobehavioral syndromes, ranging from the PVS to the postconcussion syndrome. Recent evidence documenting predominant cholinergic system damage in TBI (Salmond et al. 2005) suggests that cholinergic medications may potentiate the function of remaining basal forebrain neurons as they travel to the neocortex and hippocampus through relevant white matter tracts (Arciniegas et al. 1999).

Neoplasms have been considered less than optimal for investigating brain-behavior relationships. Brain tumors are often not clearly localizable because of complicating factors such as their wide extent, mass effect, and associated edema. However, with the help of improved neuroimaging, the neurobehavioral effects of many cerebral neoplasms can be usefully studied with detailed clinical-neuropathological correlations (Filley and Kleinschmidt-DeMasters 1995). Tumors can be particularly illustrative in terms of their effects on white matter tracts. *Gliomas*, for example, are thought to originate primarily from glial progenitor cells in the white matter (Canoll and Goldman 2008), and they spread via white matter tracts to other regions of the brain (Giese and Westphal 1996). An outstanding example of a white matter tumor producing neurobehavioral dysfunction can be seen with *gliomatosis cerebri*, a diffusely infiltrative astrocytic malignancy with a clear predilection for cerebral white matter (Filley et al. 2003). Neurobehavioral features are the most common presenting and persistent clinical manifestations of this rare tumor, which can be seen by MRI to spread via intra- and interhemispheric white matter tracts and produce progressive dementia (Filley et al. 2003). Gliomatosis cerebri illustrates once again that selective white matter dysfunction results in frequent and clinically significant cognitive and emotional disturbances (Filley et al. 2003). *Cerebral lymphoma* may demonstrate a similar clinical propensity when it takes the diffusely infiltrative form of lymphomatosis cerebri (Rollins et al. 2005). Study of brain tumors as the agents of mental status changes related to white matter dysfunction deserves more attention, especially as more powerful neuroimaging modalities offer the opportunity to view the location and spread of these tumors throughout their clinical course.

Normal pressure hydrocephalus (NPH) is a condition of dementia, gait disturbance, and incontinence that can occasionally be traced to a previous history of

TBI, meningitis, or subarachnoid hemorrhage, but is usually idiopathic. This disorder is commonly held to be a reversible dementia, as it can sometimes be effectively treated with ventricular shunting procedures (Adams et al. 1965). Pathologic studies have revealed that the periventricular white matter bears the brunt of the damage (DiRocco et al. 1977; Del Bigio 1993), and ventricular dilation without cortical involvement is evident neuroradiologically. Clinical recovery from NPH is presumably due to reduction of periventricular white matter compression after shunting, although the potential for recovery diminishes as the duration of hydrocephalus increases (Del Bigio 1993). Other factors that limit reversibility in NPH are the frequent co-occurrence of changes related to vascular disease (Earnest et al. 1974) and AD (Bech et al. 1997), both of which are common in the age range when NPH is likely to come to clinical attention.

The diversity of conditions characterized by significant damage to cerebral white matter might imply that their neurobehavioral features are equally variable. There are, however, recurrent patterns that appear in the group as a whole (Filley et al. 1988; Filley 1998, 2001). As a first approximation, the white matter dementias have been shown to resemble the classic subcortical dementias in many respects. However, important differences can be found that distinguish these dementias from both those that occur with diseases of the cortex and those that accompany dysfunction of the subcortical gray matter (Table 12.3). These distinctions serve as general guidelines, and it is important to recall that they are most relevant to the early stages of dementia, when accurate diagnosis is most critical; as dementia advances, the clinical picture tends to become similar despite what the cause of the problem may be.

To begin, slowed information processing is typical of white matter dementia, and this characteristic pervades the entire behavioral repertoire (Filley 2001). Cognitive slowing is also seen with subcortical dementia but is not prominent in cortical dementia. Attentional dysfunction, particularly involving sustained attention, appears to be more prominent in both white matter and subcortical dementia than in the cortical dementias. In the realm of memory, two sets of observations are relevant. First, a retrieval deficit characterizes the memory loss of both non-cortical dementias, in contrast to the learning deficit (amnesia) seen in cortical dementia (Filley et al. 1989; Lafosse et al. 2007). Second, recent data have shown that procedural memory is affected only in subcortical dementia (Lafosse et al. 2007). Language is usually preserved in dementias that spare the cortex, but speech is not; palilalia and hypophonia are often seen in subcortical gray matter dementias, and dysarthria is common in white matter dementias. Visuospatial skills are likely to be impaired in all varieties of dementia; cortical disease causes a conceptual deficit we have termed constructional apraxia, and subcortical gray and white matter disorders may produce impairment based on motor dysfunction. Executive function is also impaired universally, but the

TABLE 12.3. Clinical differentiation of dementia syndromes

	Cortical	Subcortical	White Matter
Information processing speed	Normal	Slow	Slow
Sustained attention	Normal	Poor	Poor
Declarative memory	Amnesia	Retrieval deficit	Retrieval deficit
Procedural memory	Normal	Poor	Normal
Speech	Normal	Hypophonia Palilalia	Dysarthria
Language	Aphasia	Normal	Normal
Visuospatial function	Constructional apraxia	Visuomotor impairment	Visuomotor impairment
Executive function	Poor	Slowing Inattention	Slowing Inattention
Emotion and personality	Apathy Irritability	Depression Psychosis	Depression Psychosis
Motor function	Apraxia	Extrapyramidal signs	Corticospinal signs Cerebellar signs
Sensory function	Agnosia	Normal	Blindness Deafness Hypesthesia

executive dysfunction seen in subcortical and white matter dementias may be heavily influenced by cognitive slowing and inattention. Emotional and personality changes differ in that the cortical dementias tend to have milder manifestations such as apathy or irritability, whereas the other categories frequently display severe depression of frank psychosis. Finally, abnormalities of motor and sensory function are distinguishable. Cortical dementia often shows apraxia, in contrast to extrapyramidal or corticospinal and cerebellar signs, and agnosia, as opposed to normal sensory function or dysfunction of primary sensory input.

MIXED DEMENTIAS

Few systems of classification are flawless, and in our consideration of dementia, several diseases merit discussion requiring the establishment of another category. Some disorders do not exclusively or even primarily affect cortical gray, subcortical gray, or cerebral white matter, and their neuropathology is diverse enough to produce quite variable clinical manifestations. Thus the category of mixed dementia will be considered.

One well-recognized member of this group is *multi-infarct dementia* (MID), a type of vascular dementia in which cerebral infarction may damage any component of the brain and result in a wide variety of deficits that defy classification within a single dementia category (Román et al. 1993). MID classically presents with a step-wise progression of acute cerebral insults, the cumulative effect of which is advancing dementia, and is regarded by many as the second most common cause of dementia after AD (Mendez and Cummings 2003). Hypertension, diabetes mellitus, heart disease, obesity, smoking, and hypercholesterolemia all predispose to this dementia, although preventive measures for these problems as well as the availability of thrombolysis with tissue plasminogen activator (tPA) present effective treatment options for many potential or actual stroke patients. The neurobehavioral features of MID are predictably variable, depending on the brain areas involved. Occasionally, dementia may even result from a single infarct in a particularly eloquent cerebral area, such as the left angular gyrus or the thalamus, producing the syndrome of *strategic infarct dementia* (Mendez and Cummings 2003). Another form of vascular dementia involving multiple infarcts is the *lacunar state*, in which multiple small, subfrontal gray and white matter ischemic lesions known as lacunes produce a dementia syndrome with prominent frontal lobe features (Ishii, Nishahara, and Imamura 1986; Chui 2007). The lacunar state shares much in common with BD but often involves lacunes in the subcortical gray matter regions as well as in white matter tracts.

Creutzfeldt-Jakob disease (CJD) is a rapidly progressive dementia, often with prominent myoclonus and a characteristic electroencephalographic pattern of periodic high-amplitude discharges (Masters and Richardson 1978; Johnson and Gibbs 1998). CJD is one of a group of prion diseases, or transmissible spongiform encephalopathies, in which an abnormal form of a normal cellular protein accumulates in the brain (Knight and Will 2004). In CJD, disease is caused either by prion infection or, in 10–15 percent of patients, by a mutation in the prion protein gene (Prusiner 1993). The disease involves spongiform degeneration and neuronal loss in the cortex, basal ganglia, thalamus, and brainstem and occasionally axonal damage and demyelination in the cerebral white matter (Johnson and Gibbs 1998). MRI has proven helpful in diagnosis, as diffusion sequences have disclosed high signal in gray matter of the cortex and basal ganglia (Yee et al. 1999), and the 14-3-3 protein in the cerebrospinal fluid adds additional value; these advances have led to a diminished need for cerebral biopsy to confirm the diagnosis of CJD (Knight and Will 2004). No treatment to reverse or stabilize CJD is available, and the disease is fatal within a year of onset in most cases. In the 1990s, a variant of CJD—that first appeared in cows as bovine spongiform encephalopathy—was observed in humans and called variant CJD (Knight and Will 2004). A human form of "mad cow disease," this disease is transmitted by the ingestion of infected beef and, in contrast to typical CJD, involves somewhat younger people and features more psychiatric manifestations (Johnson and Gibbs 1998). With containment of the infected cattle by public health measures, this variant has not persisted, and typical CJD remains the most important prion disease for humans.

Dementia with Lewy bodies (DLB) is a degenerative disorder related to PD but characterized by Lewy bodies distributed throughout cortical and subcortical regions (Ferman and Boeve 2007). First described as diffuse Lewy body disease (DLBD; Burkhardt et al. 1988), the term DLB subsequently came to be preferred as it was recognized that whereas Lewy bodies are characteristic, variable numbers of neuritic plaques and neurofibrillary tangles are also seen in the brains of affected patients (McKeith et al. 2004). While cases of DLBD (Burkhardt et al. 1988) still appear, they are rare since AD changes accompany Lewy bodies in most cases. The widespread distribution of Lewy bodies explains why DLB manifests dementia and prominent visual hallucinations in addition to the expected parkinsonism as clinical hallmarks (McKeith et al. 2004). A fluctuating confusional state has been observed in many patients, and treatment may be difficult because medications for parkinsonism may worsen psychosis while drugs for psychosis may exacerbate parkinsonism. A relatively new addition to the list of dementias, DLB has been found by some to be more common than MID, exceeded in prevalence among the dementias only by AD (Ferman and Boeve 2007).

Corticobasal degeneration is an idiopathic neurodegenerative disease that features progressive asymmetric rigidity and apraxia in association with dementia that may be relatively mild (Boeve, Lang, and Litvan 2003). Often distinctly unilateral or more evident on one side of the body, this disease is associated with other clinical features that may include aphasia, hemineglect, dystonia, myoclonus, tremor, and the alien hand sign. The unusual combination of cortical and basal ganglia dysfunction that lateralizes to one side marks this disorder as unique among the neurodegenerative diseases, although similarities to FTD and PSP have been noted (Boeve, Lang, and Litvan 2003).

Neurosyphilis is uncommonly seen in the present era but was formerly one of the most prevalent causes of dementia (Timmermans and Carr 2004). Before the introduction of penicillin in the mid-twentieth century, more than 20 percent of patients in mental hospitals had tertiary syphilis (Hutto 2001), typically in the form known as general paresis that required psychiatric treatment because of grandiose delusions and expansive mania (Mendez and Cummings 2003). The causative spirochete *Treponema pallidum* has a predilection for the frontal cortices, and this feature likely contributes to this neuropsychiatric presentation of the disease. However, many other patients have a "simple" dementia without prominent psychosis, presumably due to the disease involving nonfrontal areas (Mendez and Cummings 2003). Other forms of neurosyphilis may also result in dementia, notably meningovascular syphilis, in which multiple strokes occur in the setting of cerebral vasculitis and meningitis, and syphilitic gumma, a mass lesion that resembles other masses in its propensity to compress the brain

(Mendez and Cummings 2003). Neurosyphilis is often associated with HIV infection (Simon 1985) and is more severe and resistant to treatment in these cases (Timmermans and Carr 2004).

Mass lesions including *subdural hematoma* (Karnath 2004), which are often unassociated with significant head trauma in the elderly, and brain *neoplasms* (Taphoorn and Klein 2004) affect the cerebrum diffusely, especially if there is an associated increase in intracranial pressure. The dementia syndrome resulting from these lesions can include many diverse neurologic signs and neurobehavioral alterations, depending on the areas primarily and secondarily involved. Hemiparesis, hemihypesthesia, visual field deficits, and papilledema may be prominent in patients with these kinds of lesions. Other clinical features, such as headache and seizures, are common as well and may lead to clinical recognition before a neurologist is consulted. Brain tumors were considered above in the context of white matter dementia, but these masses may also involve gray matter; in general, regardless of the tissue involved, neurobehavioral and neuropsychiatric presentations imply a frontal or temporal location of the mass (Filley and Kleinschmidt-DeMasters 1995). It should also be noted that severe leukotoxicity can result from brain tumor therapy, mainly radiation and many chemotherapeutic agents, and produce dementia even as the neoplasm is being treated (Filley and Kleinschmidt-DeMasters 2003). Subdural hematomas and neoplasms, especially benign tumors such as meningioma, represent potentially reversible causes of dementia, especially if discovered early in their course (Hunter, Blackwood, and Bull 1968). As discussed above, brain malignancies beginning in the white matter can be readily detected by MRI, so the early diagnostic use of this modality can reveal these tumors at a time when treatment may be most efficacious.

Cerebral *hypoxia-ischemia* can result from inadequate blood oxygenation, insufficient cerebral perfusion, or anemia with reduced oxygen-carrying capacity (Mendez and Cummings 2003; Anderson and Arciniegas 2010). Although chronic pulmonary and cardiac disease can each cause a dementia syndrome due to hypoxia-ischemia, more common and dramatic is the acute onset of hypoxic-ischemic brain injury following cardiopulmonary arrest, carbon monoxide poisoning, strangulation, or anesthetic accidents (Mendez and Cummings 2003). One clinical scenario often confronting behavioral neurologists is the prognostication of outcome in severely impaired patients who are resuscitated after cardiac arrest. Profound hypoxia-ischemia can of course lead to brain death, but lesser degrees of injury can result in the PVS (Multi-Society Task Force on PVS 1994), the MCS (Giacino et al. 2002), or severe dementia (Richardson, Chambers, and Heywood 1959). Diffuse laminar cortical necrosis, usually also involving the hippocampus, is common in hypoxia-ischemia (Dougherty et al. 1981), and delayed cerebral demyelination developing days to weeks after the insult is occasionally seen (Plum, Posner, and Hain 1962; Shprecher and Mehta 2010). The long-term care of people who have sustained such devastating injuries poses a major challenge to the health-care system.

REFERENCES

Adams, J. H., Graham, D. I., Murray, L. S., and Scott, G. Diffuse axonal injury due to non-missile head injury: an analysis of 45 cases. *Ann Neurol* 1982; 12: 557–563.

Adams, R. D., Fisher, C. M., Hakim, S., et al. Symptomatic occult hydrocephalus with "normal" cerebrospinal fluid pressure: a treatable syndrome. *N Engl J Med* 1965; 273: 117–126.

Adams, R. D., and Kubik, C. S. Subacute degeneration of the brain in pernicious anemia. *N Engl J Med* 1944; 231: 1–9.

Albert, M. L. Subcortical dementia. In: Katzman, R., Terry, R. D., and Bick, K. L., eds. *Alzheimer's Disease: Senile Dementia and Related Disorders*. New York: Raven Press; 1978: 173–180.

Albert, M. L., Feldman, R. G., and Willis, A. L. The "subcortical dementia" of progressive supranuclear palsy. *J Neurol Neurosurg Psychiatry* 1974; 37: 121–130.

Alexander, M. P. Mild traumatic brain injury: pathophysiology, natural history, and clinical management. *Neurology* 1995: 45: 1252–1260.

Almkvist, O., Wahlund, L.-O., Andersson-Lundman, G., et al. White-matter hyperintensity and neuropsychological functions in dementia and healthy aging. *Arch Neurol* 1992; 49: 626–632.

Alzenstein, H. J., Nebes, R. D., Saxton, J. A., et al. Frequent amyloid deposition without significant cognitive impairment among the elderly. *Arch Neurol* 2008; 65: 1509–1517.

Alzheimer, A. Über eine eigenartige Erkrankung der Hirnrinde. *Allgemeine Zeitschrift für Psychiatrie und Psychisch-Gerichtliche Medizin* 1907; 64: 146–148.

American Psychiatric Association. *Diagnostic and Statistical Manual of Mental Disorders*. 4th ed. Washington, DC: American Psychiatric Association; 1994.

Anderson, C. A., and Arciniegas, D. B. Cognitive sequelae of hypoxic-ischemic brain injury: a review. *NeuroRehabilitation* 2010; 26: 47–63.

Arciniegas, D., Adler, L., Topkoff, J., et al. Attention and memory dysfunction after traumatic brain injury: cholinergic mechanisms, sensory gating, and a hypothesis for further investigation. *Brain Injury* 1999; 13: 1–13.

Austin, J., Armstrong, D., Fouch, S., et al. Metachromatic leukodystrophy (MLD); vol. VIII: MLD in adults: diagnosis and pathogenesis. *Arch Neurol* 1968; 18: 225–240.

Babikian, V., and Ropper, A. H. Binswanger's disease: a review. *Stroke* 1987; 18: 2–12.

Bartzokis, G. Alzheimer's disease as homeostatic responses to age-related myelin breakdown. *Neurobiol Aging* 2009; September 21 [e-pub ahead of print].

Bearden, C. E., Hoffman, K. M., and Cannon, T. D. The neuropsychology and neuroanatomy of bipolar affective disorder: a critical review. *Bipolar Disord* 2001; 3: 106–150.

Bech, R. A., Juhler, M., Waldemar, G., et al. Frontal brain and leptomeningeal biopsy specimens correlated with cerebrospinal fluid outflow resistance and B-wave activity in patients suspected of normal-pressure hydrocephalus. *Neurosurgery* 1997; 40: 497–502.

Bigio, E. H. Update on recent molecular and genetic advances in frontotemporal lobar degeneration. *J Neuropathol Exp Neurol* 2008; 67: 635–648.

Binswanger, O. Die Abgrenzung der allgemeinen progressiven Paralyse. *Berl Klin Wochenschr* 1894; 31: 1102–1105, 1137–1139, 1180–1186.

Blessed, G., Tomlinson, B. E., and Roth, M. The association between quantitative measures of dementia and of senile change in the cerebral grey matter of elderly subjects. *Br J Psychiatry* 1968; 114: 797–811.

Boerner, R. J., and Kapfhammer, H. P. Psychopathological changes and cognitive impairment in encephalomyelitis disseminata. *Eur Arch Clin Neurosci* 1999; 249: 96–102.

Boeve, B. F., Lang, A. E., and Litvan, I. Corticobasal degeneration and its relationship to progressive supranuclear palsy and frontotemporal dementia. *Ann Neurol* 2003: 54 (Suppl 5): S15–S19.

Boissé, L., Gill, M. J., and Power, C. HIV infection of the central nervous system: clinical features and neuropathogenesis. *Neurol Clin* 2008; 26: 799–819.

Bonelli, R. M., and Cummings, J. L. Frontal-subcortical dementias. *Neurologist* 2008; 14: 100–107.

Boone, K., Miller, B. L., Lesser, I. M., et al. Neuropsychological correlates of white-matter lesions in healthy elderly subjects: a threshold effect. *Arch Neurol* 1992; 49: 549–554.

Brandt, J., Folstein, S. E., and Folstein, M. F. Differential cognitive impairment in Alzheimer's disease and Huntington's disease. *Ann Neurol* 1988; 23: 555–561.

Budka, H. Neuropathology of human immunodeficiency virus infection. *Brain Pathol* 1991; 1: 163–175.

Burkhardt, C. R., Filley, C. M., Kleinschmidt-DeMasters, B. K., et al. Diffuse Lewy body disease and progressive dementia. *Neurology* 1988; 38: 1520–1528.

Caine, E. D. Pseudodementia: current concepts and future directions. *Arch Gen Psychiatry* 1981; 38: 1359–1364.

Cairns, N. J., and Ghoshal, N. FUS: a new actor on the frontotemporal dementia stage. *Neurology* 2010; 74: 354–356.

Canoll, P., and Goldman, J. E. The interface between glial progenitors and gliomas. *Acta Neuropathol* 2008; 116: 465–477.

Caplan, L. R. Binswanger's disease—revisited. *Neurology* 1995; 45: 626–633.

Carlen, P. L., and Wilkinson, D. A. Reversibility of alcohol-related brain damage: clinical and experimental observations. *Acta Med Scand Suppl* 1987; 717: 19–26.

Charcot, J. M. *Lectures on the Diseases of the Nervous System Delivered at La Salpêtrière.* London: New Sydenham Society; 1877.

Chatterjee, A., Yapundich, R., Palmer, C. A., et al. Leukoencephalopathy associated with cobalamin deficiency. *Neurology* 1996; 46: 832–834.

Chui, H. Subcortical ischemic vascular dementia. *Neurol Clin* 2007; 25: 717–740.

Clark, R. F., and Goate, A. M. Molecular genetics of Alzheimer's disease. *Arch Neurol* 1993; 50: 1164–1172.

Corder, E. H., Saunders, A. M., Strittmatter, W. J., et al. Gene dose of apolipoprotein E type 4 allele and the risk of Alzheimer's disease in late onset families. *Science* 1993; 261: 921–923.

Crook, T., Bartus, R. T., Ferris, S. H., et al. Age-associated memory impairment: proposed diagnostic criteria and measures of clinical change. Report of a National Institute of Mental Health work group. *Dev Neuropsychol* 1986; 2: 261–276.

Cullum, C. M., Butters, N., Troster, A. I., and Salmon, D. P. Normal aging and forgetting rates on the Wechsler Memory Scale—Revised. *Arch Clin Neuropsychol* 1990; 5: 23–30.

Cummings, J. L. Alzheimer's disease. *N Engl J Med* 2004; 351: 56–67.

———. Intellectual impairment in Parkinson's disease: clinical, pathologic, and biochemical correlates. *J Geriatr Psychiatry Neurol* 1988; 1: 24–36.

———. The neuroanatomy of depression. *J Clin Psychiatry* 1993; 54 (Suppl): 14–20.

———. Organic psychoses: delusional disorders and secondary mania. *Psychiat Clin N Am* 1986; 9: 293–311.

Cummings, J. L., and Benson, D. F. Subcortical dementia: review of an emerging concept. *Arch Neurol* 1984; 41: 874–879.

Cummings, J. L., and Mega, M. S. *Neuropsychiatry and Behavioral Neuroscience*. New York: Oxford University Press; 2003.

De la Monte, S. Disproportionate atrophy of cerebral white matter in chronic alcoholics. *Arch Neurol* 1988; 45: 990–992.

Del Bigio, M. R. Neuropathological changes caused by hydrocephalus. *Acta Neuropathol* 1993; 85: 573–585.

DiRocco, C., DiTrapani, G., Maira, G., et al. Anatomo-clinical correlations in normotensive hydrocephalus. *J Neurol Sci* 1977; 33: 437–452.

Dougherty, J. H., Rawlinson, D. G., Levy, D. E., and Plum, F. Hypoxic-ischemic brain injury and the vegetative state: clinical and neuropathologic correlations. *Neurology* 1981; 31: 991–997.

Drebing, C. E., Moore, L. H., Cummings, J. L., et al. Patterns of neuropsychological performance among forms of subcortical dementia. *Neuropsychiatry Neuropsychol Behav Neurol* 1994; 7: 57–66.

Earnest, M. P., Fahn, S., Karp, J. H., and Rowland, L. P. Normal pressure hydrocephalus and hypertensive cerebrovascular disease. *Arch Neurol* 1974; 31: 262–266.

Ferman, T. J., and Boeve, B. Dementia with Lewy bodies. *Neurol Clin* 2007; 25: 741–760.

Filley, C. M. The behavioral neurology of cerebral white matter. *Neurology* 1998; 50: 1535–1540.

———. *The Behavioral Neurology of White Matter*. New York: Oxford University Press; 2001.

———. White matter and behavioral neurology. *Ann NY Acad Sci* 2005; 1064: 162–183.

Filley C. M., and Cullum, C. M. Attention and vigilance functions in normal aging. *Appl Neuropsychol* 1994; 1: 29–32.

———. Early detection of frontal-temporal degeneration by clinical evaluation. *Arch Clin Neuropsychol* 1993; 8: 359–367.

———. Education and cognitive function in Alzheimer's disease. *Neuropsychiatry Neuropsychol Behav Neurol* 1997; 10: 48–51.

Filley, C. M., Davis, K. A., Schmitz, S. P., et al. Neuropsychological performance and magnetic resonance imaging in Alzheimer's disease and normal aging. *Neuropsychiatry Neuropsychol Behav Neurol* 1989; 2: 81–91.

Filley, C. M., Franklin, G. M., Heaton, R. K., and Rosenberg, N. L. White matter dementia: clinical disorders and implications. *Neuropsychiatry Neuropsychol Behav Neurol* 1988; 1: 239–254.

Filley, C. M., and Gross, K. F. Psychosis with cerebral white matter disease. *Neuropsychiatry Neuropsychol Behav Neurol* 1992; 5: 119–125.

Filley, C. M., Heaton, R. K., Nelson, L. M., et al. A comparison of dementia in Alzheimer's disease and multiple sclerosis. *Arch Neurol* 1989; 46: 157–161.

Filley, C. M., Heaton, R. K., and Rosenberg, N. L. White matter dementia in chronic toluene abuse. *Neurology* 1990; 40: 532–534.

Filley, C. M., and Kleinschmidt-DeMasters, B. K. Neurobehavioral presentations of brain neoplasms. *West J Med* 1995; 163: 19–25.

———. Toxic leukoencephalopathy. *N Engl J Med* 2003; 345: 425–432.

Filley, C. M., Kleinschmidt-DeMasters, B. K., and Gross, K. F. Non-Alzheimer frontotemporal degenerative dementia: a neurobehavioral and pathologic study. *Clin Neuropathol* 1994; 13: 109–116.

Filley, C. M., Kleinschmidt-DeMasters, B. K., Lillehei, K. O., et al. Gliomatosis cerebri: neurobehavioral and neuropathological observations. *Cogn Behav Neurol* 2003; 16: 149–159.

Filley, C. M., Kozora, E., Brown, M. S., et al. White matter microstructure and cognition in non-neuropsychiatric systemic lupus erythematosus. *Cogn Behav Neurol* 2009; 22: 38–44.

Folstein, M. F., Folstein, S. E., and McHugh, P. R. "Mini-mental state": a practical method for grading the cognitive state of patients for the clinician. *J Psychiat Res* 1975; 12: 189–198.

Franklin, G. M., Heaton, R. K., Nelson, L. M., et al. Correlation of neuropsychological and magnetic resonance imaging findings in chronic/progressive multiple sclerosis. *Neurology* 1988; 38: 1826–1829.

Gazdzinski, S., Durazzo, T. C., Mon, A., et al. Cerebral white matter recovery in abstinent alcoholics: a multimodality magnetic resonance study. *Brain* 2010; 133: 1043–1053.

Geschwind, N. Disconnexion syndromes in animals and man. *Brain* 1965; 88: 237–294, 585–644.

Giacino, J. T., Ashwal, S., Childs, N., et al. The minimally conscious state: definition and diagnostic criteria. *Neurology* 2002; 58: 349–353.

Giese, A., and Westphal, M. Glioma invasion in the central nervous system. *Neurosurgery* 1996; 39: 235–250.

Goodkin, D. E., Rudick, R. A., and Ross, J. S. The use of brain magnetic resonance imaging in multiple sclerosis. *Arch Neurol* 1994; 51: 505–516.

Guze, B. H., and Gitlin, M. The neuropathologic basis of major affective disorders. *J Neuropsychiatry* 1994; 6: 114–121.

Hachinski, V. C., Potter, P., and Merskey, H. Leuko-araiosis. *Arch Neurol* 1987; 44: 21–23.

Hansen, L. A., Natelson, B. H., Lemere, C., et al. Alcohol-induced brain changes in dogs. *Arch Neurol* 1991; 48: 939–942.

Hardy, J. The amyloid hypothesis for Alzheimer's disease: a critical reappraisal. *J Neurochem* 2009; 110: 1129–1134.

Heaton, R. K., Nelson, L. M., Thompson, D. S., et al. Neuropsychological findings in relapsing-remitting and chronic-progressive multiple sclerosis. *J Consul Clin Psychol* 1985; 53: 103–110.

Hormes, J. T., Filley, C. M., and Rosenberg, N. L. Neurologic sequelae of chronic solvent vapor abuse. *Neurology* 1986; 36: 698–672.

Howard, K. L., and Filley, C. M. Advances in genetic testing for Alzheimer's disease. *Rev Neurol Dis* 2009; 6: 26–32.

Huber, S. J., and Shuttleworth, E. C. Neuropsychological assessment of subcortical dementia. In: Cummings, J. L., ed. *Subcortical Dementia*. New York: Oxford; 1990: 71–86.

Huber, S. J., Shuttleworth, E. C., Paulson, G. W., et al. Cortical vs. subcortical dementia: neuropsychological differences. *Arch Neurol* 1986; 43: 392–394.

Hunter, R., Blackwood, W., and Bull, J. Three cases of frontal meningiomas presenting psychiatrically. *Br Med J* 1968; 3: 9–16.

Hutto, B. Syphilis in clinical psychiatry: a review. *Psychosomatics* 2001; 42: 453–460.

Hyde, T. M., Ziegler, J. C., and Weinberger, D. R. Psychiatric disturbances in metachromatic leukodystrophy: insights into the neurobiology of psychosis. *Arch Neurol* 1992; 49: 401–406.

Ishii, N., Nishahara, Y., and Imamura, T. Why do frontal lobe symptoms predominate in vascular dementia with lacunes? *Neurology* 1986; 36: 340–345.

Jensen, G. B., and Pakkenberg, B. Do alcoholics drink their neurons away? *Lancet* 1993; 342: 1201–1204.

Jeste, D. V., Lohr, J. B., and Goodwin, F. K. Neuroanatomical studies of major affective disorders: a review and suggestions for further research. *Br J Psychiatry* 1988; 153: 444–459.

Johnson, R. T., and Gibbs, G. J. Creutzfeldt-Jakob disease and related transmissible spongiform encephalopathies. *N Engl J Med* 1998; 339: 1994–2004.

Josephs, K. Frontotemporal lobar degeneration. *Neurol Clin* 2007; 25: 683–696.

Junque, C., Pujol, J., Vendrell, P., et al. Leuko-araiosis on magnetic resonance imaging and speed of mental processing. *Arch Neurol* 1990; 47: 151–156.

Karnath, B. Subdural hematoma: presentation and management in older adults. *Geriatrics* 2004; 59: 18–23.

Katzman, R. Alzheimer's disease. *N Engl J Med* 1986; 314: 964–973.

———. The prevalence and malignancy of Alzheimer disease. *Arch Neurol* 1976; 33: 217–218.

King, D. A., and Caine, E. D. Depression. In: Cummings, J. L., ed. *Subcortical Dementia*. New York: Oxford University Press; 1990: 218–230.

Kirk, A., Kertesz, A., and Polk, M. J. Dementia with leukoencephalopathy in systemic lupus erythematosus. *Can J Neurol Sci* 1991; 18: 344–348.

Knight, R.S.G., and Will, R. G. Prion diseases. *J Neurol Neurosurg Psychiatry* 2004; 75 (Suppl 1): 36–42.

Koenigs, M., and Grafman, J. The functional neuroanatomy of depression: distinct roles for ventromedial and dorsolateral prefrontal cortex. *Behav Brain Res* 2009 12; 201: 239–243.

Kozora, E., Thompson, L. L., West, S. G., and Kotzin, B. L. Analysis of cognitive and psychological deficits in systemic lupus erythematosus patients without overt central nervous system disease. *Arthritis Rheum* 1996; 39: 2035–2045.

Kral, V. A. Senescent forgetfulness: benign and malignant. *Can Med Assoc J* 1962; 86: 257–260.

Lafosse, J. M., Corboy, J. R., Leehey, M. A., et al. MS vs. HD: can white matter and subcortical gray matter pathology be distinguished neuropsychologically? *J Clin Exp Neuropsychol* 2007; 29: 142–154.

Lindenbaum, J., Healton, E. B., Savage, D. G., et al. Neuropsychiatric disorders caused by cobalamin deficiency in the absence of anemia or macrocytosis. *N Engl J Med* 1988; 318: 1720–1728.

Lishman, W. A. Cerebral disorder in alcoholism. *Brain* 1981; 104: 1–20.

Marx, J. Preventing Alzheimer's: a lifelong commitment? *Science* 2005; 309: 864–866.

Masters, C. L., and Richardson, E. P. Subacute spongiform encephalopathy (Creutzfeldt-Jakob disease). *Brain* 1978; 101: 333–344.

McHugh, P. R., and Folstein, M. E. Psychiatric syndromes of Huntington's chorea: a clinical and phenomenologic study. In: Benson, D. F., and Blumer, D., eds. *Psychiatric Aspects of Neurologic Disease*. Vol. 1. New York: Grune and Stratton; 1975: 267–285.

McKeith, I., Mintzer, J., Aarsland, D., et al. Dementia with Lewy bodies. *Lancet Neurology* 2004; 3: 19–28.

Medalia, A., Isaacs-Glaberman, K., and Scheinberg, I. H. Neuropsychological impairment in Wilson's disease. *Arch Neurol* 1988; 45: 502–504.

Mendez, M. F. The neurobiology of moral behavior: review and neuropsychiatric implications. *CNS Spectr* 2009; 14: 608–620.

Mendez, M. F., and Cummings, J. L. *Dementia: A Clinical Approach*. 3rd ed. Philadelphia: Butterworth-Heinemann; 2003.

Mesulam, M.-M. Primary progressive aphasia: a 25-year retrospective. *Alzheimer Dis Assoc Disord* 2007; 21: S8–S11.

Miller, A.K.H., Alston, R. L., and Corsellis, J.A.N. Variation with age in the volumes of grey and white matter in the cerebral hemispheres of man: measurements with an image analyzer. *Neuropathol Appl Neurobiol* 1980; 6: 119–132.

Multi-Society Task Force on PVS. Medical aspects of the persistent vegetative state. *N Engl J Med* 1994; 330: 1499–1508, 1572–1579.

Navia, B. A., Cho, E. S., Petito, C. K., and Price, R. W. The AIDS dementia complex; II: Neuropathology. *Ann Neurol* 1986; 19: 525–535.

Navia, B. A., Jordan, B. D., and Price, R. W. The AIDS dementia complex; I: Clinical features. *Ann Neurol* 1986; 19: 517–524.

Nolte, J. *The Human Brain: An Introduction to Its Functional Anatomy*. St. Louis: Mosby; 2002.

Osuji, I. J., and Cullum, C. M. Cognition in bipolar disorder. *Psychiatr Clin North Am* 2005; 28: 427–441.

Parker, W. D., Parks, J., Filley, C. M., and Kleinschmidt-DeMasters, B. K. Electron transport chain defects in Alzheimer's disease brain. *Neurology* 1994; 44: 1090–1096.

Paskavitz, J. F., Anderson, C. A., Filley C. M., et al. Acute arcuate fiber demyelinating encephalopathy following Epstein-Barr virus infection. *Ann Neurol* 1995; 38: 127–131.

Perl, D. P., and Brody, A. R. Alzheimer's disease: x-ray spectrometric evidence of aluminum bearing accumulation in neurofibrillary tangle bearing neurons. *Science* 1980; 208: 297–299.

Petersen, R. C., and Negash, S. Mild cognitive impairment: an overview. *CNS Spectr* 2008; 13: 45–53.

Petersen, R. C., Smith, G. E., Waring, S. C., et al. Mild cognitive impairment: clinical characterization and outcome. *Arch Neurol* 1999; 56: 303–308.

Pick, A. Über die Beziehungen der Senilin Hirnatrophie zur Aphasie. *Prager Med Wochenschr* 1892; 17: 165–167.

Pillon, B., Dubois, B., Lhermitte, F., and Agid, Y. Heterogeneity of cognitive impairment in progressive supranuclear palsy, Parkinson's disease, and Alzheimer's disease. *Neurology* 1986; 36: 1179–1185.

Plum, F., Posner, J. B., and Hain, R. F. Delayed neurological deterioration after anoxia. *Arch Int Med* 1962; 110: 18–25.

Price, R. W., Brew, B., Sidtis, J., et al. The brain in AIDS: central nervous system HIV-1 infection and AIDS dementia complex. *Science* 1988; 239: 586–592.

Prusiner, S. B. Genetic and infectious prion diseases. *Arch Neurol* 1993; 50: 1129–1153.

Querfurth, H. W., and LaFerla, F. M. Alzheimer's disease. *N Engl J Med* 2010; 362: 329–344.

Rao, S. M., Leo, G. J., Bernardin, L., and Unverzagt, F. Cognitive dysfunction in multiple sclerosis; I: Frequency, patterns, and prediction. *Neurology* 1991; 41: 685–691.

Richardson, J. C., Chambers, R. A., and Heywood, P. M. Encephalopathies of anoxia and hypoglycemia. *Arch Neurol* 1959; 1: 178–190.

Rollins, K. E., Kleinschmidt-DeMasters, B. K., Corboy, J. R., et al. Lymphomatosis cerebri as a cause of white matter dementia. *Hum Pathol* 2005; 36: 282–290.

Román, G. C. From UBOs to Binswanger's disease: impact of magnetic resonance imaging on vascular dementia research. *Stroke* 1996; 27: 1269–1273.

Román, G. C., Tatemichi, T. K., Erkinjuntti, T., et al. Vascular dementia: diagnostic criteria for research studies. Report of the NINDS-AIREN International Workshop. *Neurology* 1993; 43: 250–260.

Ropper, A., and Samuels, M. *Adams and Victor's Principles of Neurology.* 9th ed. New York: McGraw-Hill; 2009.

Rosen, H. J., Allison, S. C., Schauer, G. F., et al. Neuroanatomical correlates of behavioural disorders in dementia. *Brain* 2005; 128: 2612–2625.

Rosenberg, N. L., Kleinschmidt-DeMasters, B. K., Davis, K. A., et al. Toluene abuse causes diffuse central nervous system white matter changes. *Ann Neurol* 1988; 23: 611–614.

St. George-Hyslop, P. H., Tanzi, R. E., Polinsky, R. J., et al. The genetic defect causing familial Alzheimer's disease maps on chromosome 21. *Science* 1987; 235: 885–890.

Salloway, S., and Cummings, J. Subcortical disease and neuropsychiatric illness. *J Neuropsychiatry Clin Neurosci* 1994; 6: 93–99.

Salmond, C. H., Chatfield, D. A., Menon, D. K., et al. Cognitive sequelae of head injury: involvement of basal forebrain and associated structures. *Brain* 2005; 128: 189–200.

Saver, J. Time is brain—quantified. *Stroke* 2006; 37: 263–266.

Schellenberg, G. H., Bird, T. D., Wijsman, E. M., et al. Genetic linkage evidence for a familial Alzheimer's disease locus on chromosome 14. *Science* 1992; 258: 668–671.

Schmahmann, J. D., and Pandya, D. *Fiber Pathways of the Brain.* New York: Oxford University Press; 2006.

Schmidt, R., Fazekas, F., Offenbacher, H., et al. Neuropsychologic correlates of MRI white matter hyperintensities: a study of 150 normal volunteers. *Neurology* 1993; 43: 2490–2494.

Schneider, J. A., Arvanitakis, Z., Leurgans, S. E., and Bennett D. A. The neuropathology of probable Alzheimer disease and mild cognitive impairment. *Ann Neurol* 2009; 66: 200–208.

Schoenemann, P. T., Sheehan, M. J., and Glotzer, I. D. Prefrontal white matter volume is disproportionately larger in humans than in other primates. *Nat Neurosci* 2005; 8: 242–254.

Seilhean, D., Duyckaerts, C., Vazeax, R., et al. HIV-1-associated cognitive/motor complex: absence of neuronal loss in the cerebral neocortex. *Neurology* 1993; 43: 1492–1499.

Selnes, O. A., and Vinters, H. V. Vascular cognitive impairment. *Nat Clin Prac Neurol* 2006; 2: 538–547.

Seymour, S. E., Reuter-Lorenz, P. A., and Gazzaniga, M. S. The disconnection syndrome: basic findings reaffirmed. *Brain* 1994; 117: 105–115.

Shapiro, E. G., Lockman, L. A., Knopman, D., and Krivit, W. Characteristics of the dementia in late-onset metachromatic leukodystrophy. *Neurology* 1994; 44: 662–665.

Shprecher, D., and Mehta, L. The syndrome of delayed post-hypoxic leukoencephalopathy. *NeuroRehabilitation* 2010; 26: 65–72.

Simon, R. P. Neurosyphilis. *Arch Neurol* 1985; 42: 606–613.

Sollberger, M., Stanley, C. M., Wilson, S. M., et al. Neural basis of interpersonal traits in neurodegenerative diseases. *Neuropsychologia* 2009; 47: 2812–2827.

Stadelmann, C., Albert, M., Wegner, C., and Brück, W. Cortical pathology in multiple sclerosis. *Curr Opin Neurol* 2008; 21: 229–234.

Starkstein, S. E., Mayberg, H. S., Berthier, M. L., et al. Mania after brain injury: neuroradiological and metabolic findings. *Ann Neurol* 1990; 27: 652–659.

Stern, Y. Cognitive reserve. *Neuropsychologia* 2009; 47: 2015–2028.

Stojsavljević, N., Lević, Z., Drulović, J., and Dragutinović, G. A 44-month clinical-brain MRI follow-up in a patient with B_{12} deficiency. *Neurology* 1997; 49: 878–881.

Swerdlow, R. H., and Khan, S. M. The Alzheimer's disease mitochondrial cascade hypothesis: an update. *Exp Neurol* 2009; 218: 308–315.

Taphoorn, M. J., and Klein, M. Cognitive deficits in adult patients with brain tumours. *Lancet Neurol* 2004; 3: 159–168.

Terry, R. D., Masliah, E., Salmon, D. P., et al. Physical basis of cognitive alterations in Alzheimer's disease: synapse loss is the major correlate of cognitive impairment. *Ann Neurol* 1991; 30: 572–580.

Thurnher, M. M., Schindler, E. G., Thurnher, S. A., et al. Highly active antiretroviral therapy for patients with AIDS dementia complex: effect on MR imaging findings and clinical course. *AJNR* 2000; 21: 670–678.

Timmermans, M., and Carr, J. Neurosyphilis in the modern era. *J Neurol Neurosurg Psychiatry* 2004; 75: 1727–1730.

Victor, M., Adams, R. D., and Collins, G. H. *The Wernicke-Korsakoff Syndrome.* 2nd ed. Philadelphia: F. A. Davis; 1989.

Wells, C. E. Pseudodementia. *Am J Psychiatry* 1979; 136: 895–900.

Wernicke, C. *Der Aphasiche Symptomencomplex.* Breslau: Cohn and Weigert; 1874.

West, S. G. Neuropsychiatric lupus. *Rheum Dis Clin N Am* 1994; 20: 129–158.

Whitehouse, P. J. The concept of subcortical and cortical dementia: another look. *Ann Neurol* 1986; 19: 1–6.

Whitehouse, P. J., Price, D. L., Struble, R. G., et al. Alzheimer's disease and senile dementia: loss of neurons in the basal forebrain. *Science* 1982; 215: 1237–1239.

Yee, A. S., Simon, J. H., Anderson, C. A., et al. Diffusion-weighted MRI of right hemisphere dysfunction in Creutzfeldt-Jakob disease. *Neurology* 1999; 52: 1514–1515.

Ylikoski, R., Ylikoski, A., Erkinjuntti, T., et al. White matter changes in healthy elderly persons correlate with attention and speed of mental processing. *Arch Neurol* 1993; 50: 818–824.

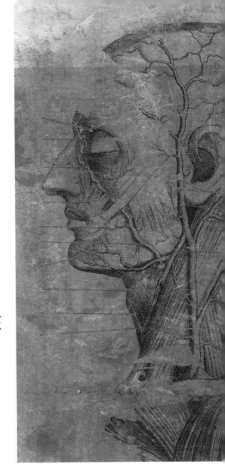

EPILOGUE

The neurology clinic, with its abundance of unfortunate but instructive lesions of the brain, offers a continuously operating laboratory for the exploration of human behavior. The clinician's first responsibility is the care of patients, but this imperative is rendered more attainable and more intelligent by attention to the ever fascinating relationships between brain and behavior. The treatment of patients in need and the understanding of the brain are both well served by the study of neurobehavioral syndromes consequent to brain disease, injury, or intoxication.

This book has been intended to describe the anatomy of higher brain function through an analysis of neurologically induced behavioral dysfunction. The syndromes of behavioral neurology have therefore formed the basis of our approach. This method is naturally constrained by the limited number and variety of patients who come to medical attention, but for the truly human capacities that characterize our species, no satisfactory substitute exists among nonhuman animals. For the direct analysis of human behavior and its disorders there is no alternative to the careful and systematic study of brain-damaged individuals.

Yet the identification of regions of the brain associated with various behaviors is only a beginning. The power of the lesion method, the elegance of sensitive new neuroimaging techniques, and the insights of neuropsychological evaluation will provide increasingly detailed understanding of the behavioral relevance of brain regions, but many other levels of analysis will be necessary as the understanding of brain-behavior relationships becomes more sophisticated. Neuropathology, which has informed all work in behavioral neurology, offers invaluable observations that cannot be acquired in any other manner. The field of neuropharmacology, upon which we have barely touched, expands the understanding of the emerging neurobehavioral architecture of the brain. Neurophysiology, to which we have made occasional reference, has an indispensable role as well, elucidating how the brain operates in parallel with studies on where its operations are localized. At still more basic levels, genetics, biochemistry, and molecular biology provide fundamental data on the origin, development, aging, recovery, and plasticity of the brain. All of these disciplines consider the broader aspects of brain neuroscience, gathering and synthesizing data from many sources to understand the brain as it engages in the activities of the mind. More than philosophy could ever accomplish alone, these many lines of inquiry will approach questions of brain and mind that can now begin to be answered by empirical data.

Even the intractable problem of consciousness may begin to yield to the combined efforts of investigators studying the intersection of brain and behavior. The assumption that consciousness is equivalent to awareness of the self and the environment may be a useful first step, but the assignment of consciousness to any single neuron, nucleus, tract, gyrus, or lobe of the brain is misguided. Can consciousness therefore be explained as an emergent property of the brain's networks working in synchrony, as the music of an orchestra arises from the concerted actions of its musicians? Although such an appealing image must for now remain metaphorical, the nature of consciousness will become steadily more comprehensible as the systematic accumulation of neuroscientific information provides a foundation for its consideration.

To conclude this brief survey of brain and behavior, then, what has been learned by clinical neuroscientists offers a compelling vista of what lies ahead. The prospects for understanding ourselves have never been brighter, and the opportunities to improve the human condition never more promising. Perhaps the sufferings of patients from whom so much has been learned can best be rendered meaningful through the knowledge gained from their misfortune.

GLOSSARY OF
NEUROBEHAVIORAL TERMS

abulia—severe apathy or loss of motivation, commonly seen with medial frontal lesions.

acalculia—an acquired disorder of calculation that often follows damage to the left parietal lobe.

acquired sociopathy—a syndrome of antisocial behavior and disinhibition resulting from bilateral orbitofrontal damage.

acute confusional state—a rapidly evolving disorder of attention (see *delirium*).

agnosia—a failure of recognition through one sensory modality; visual, auditory, and tactile agnosias have been described.

agraphesthesia—impaired recognition of numbers or letters traced on the skin of the palm.

agraphia—an acquired disorder of writing, often seen with aphasia and other neurobehavioral syndromes.

akinetic mutism—a disorder of profound abulia (see above) usually due to extensive medial bifrontal destruction.

akinetopsia—impaired movement perception, seen with lesions of the dorsal visual association cortices.

alcoholic hallucinosis—a syndrome of auditory hallucinations, often persecutory, in the setting of alcohol withdrawal.

alexia—an acquired disorder of reading; alexia with agraphia and alexia without agraphia (pure alexia) are classic syndromes.

alexia with agraphia—an acquired reading and writing disturbance due to a left angular gyrus lesion.

alexia without agraphia—an acquired reading disturbance with normal writing due to lesions in the left occipital lobe and the splenium of the corpus callosum; a synonym is pure alexia.

amnesia—an acquired disorder of recent declarative memory, implying an impairment of new learning, caused by lesions in the medial temporal lobe, diencephalon, or basal forebrain.

amusia—an acquired loss of musical skill due to focal or diffuse brain lesions, often in the right hemisphere.

anarithmetria—an isolated deficit in calculation ability, seen with left parietal lesions.

angular gyrus syndrome—Gerstmann's syndrome, anomic aphasia, and alexia with agraphia resulting from a lesion of the left angular gyrus.

anomia—impaired ability to identify objects by name, characteristic of aphasia; a synonym is dysnomia.

anosodiaphoria—awareness of but unconcern about disability, usually seen with right hemisphere lesions (see *anosognosia* and *denial*).

anosognosia—unawareness of disability, usually seen with right hemisphere lesions (see *denial* and *anosodiaphoria*).

anterograde amnesia—inability to learn new information in patients with amnesic syndromes.

Anton's syndrome—a form of anosognosia in which there is unawareness of cortical blindness.

aphasia—an acquired disorder of language, typically due to focal left hemisphere lesions; a synonym is dysphasia.

aphemia—a nonaphasic syndrome of mutism or severe dysarthria resulting from a small lesion of the lower portion of the left precentral gyrus.

apperceptive visual agnosia—failure to recognize a visual stimulus because of impaired visual perception; seen with bilateral occipital lesions.

apraxia—an impairment of learned motor activity, commonly seen with aphasia-producing lesions; classic forms are limb kinetic, ideomotor, and ideational; a synonym is dyspraxia.

aprosody—a disorder of the emotional or affective components of language, associated with focal right hemisphere lesions; a synonym is aprosodia.

associative visual agnosia—failure to recognize a visual stimulus because of inability to derive its meaning even though it is adequately perceived; seen with bilateral occipitotemporal lesions.

attention-deficit/hyperactivity disorder—a syndrome of inattention and/or hyperactivity beginning in childhood and persisting in some cases into adulthood.

Balint's syndrome—simultanagnosia, optic ataxia, and ocular apraxia (oculomotor apraxia or psychic paralysis of gaze); seen with bilateral occipitoparietal lesions.

benign senescent forgetfulness—mild memory retrieval impairment in elderly persons that does not cause the functional disability of amnesia or dementia.

blast injury—a form of traumatic brain injury in which diffuse damage results from a rapid burst of air from a nearby explosion.

blindsight—residual vision in cortically blind patients, thought to be mediated by preserved visual function in the superior colliculi, pulvinars, and parietal cortices.

Broca's aphasia—nonfluent language with relatively preserved auditory comprehension, typically due to a large lesion in the left inferior frontal lobe.

Capgras syndrome—a delusional misidentification in which a patient believes a familiar person has been replaced by an impostor; seen with bifrontal and right hemisphere lesions.

catastrophic reaction—severe depression, agitation, and hostility in some patients with severe aphasia from left hemisphere lesions.

central achromatopsia—inability to perceive colors, due to occipitotemporal lesions.

Charles Bonnet syndrome—formed visual hallucinations associated with encroaching blindness, usually resulting from ocular pathology.

chronic traumatic encephalopathy—a progressive neurodegenerative disease related to repeated sublethal traumatic brain injury in boxers and other athletes and characterized by dementia, parkinsonism, and speech and gait disorders; also known as dementia pugilistica.

color anomia—inability to name colors, due to disconnection of visual from verbal areas; typically seen with pure alexia.

coma—a disorder of the level of consciousness characterized by unarousable unresponsiveness, due to upper brainstem or extensive bihemispheric dysfunction.

conceptual apraxia—difficulty executing a learned motor act because of content errors—not understanding the concept of the action—as opposed to production errors, often seen with left parietal lesions (see *ideomotor apraxia*).

concussion—a traumatically induced alteration in mental status.

conduction aphasia—impaired repetition with intact language fluency and auditory comprehension resulting from damage of the left arcuate fasciculus or adjacent areas.

confabulation—the unintentional recitation of fictitious experiences in response to direct questioning; seen in some but not all patients with amnesia.

confusion—the inability to maintain a coherent line of thought despite adequate arousal and language function.

constructional apraxia—a term used to refer to visuospatial dysfunction (see below).

contrecoup lesion—a type of contusion involving the cerebrum directly opposite the site of impact in traumatic brain injury.

contusion—a traumatically induced bruise of the cortex and underlying white matter, typically seen in frontal polar and anterior temporal regions.

cortical blindness—blindness due to bilateral damage to the primary visual cortices or their subjacent white matter.

cortical deafness—deafness due to bilateral damage to the primary auditory cortices or their subjacent white matter.

coup lesion—a type of contusion involving the cerebrum directly underlying the cranial site of impact in traumatic brain injury.

crossed aphasia—aphasia in a right-handed patient with a right hemisphere lesion.

delirium—an agitated acute confusional state featuring delusions, hallucinations, and signs of autonomic overactivity such as tachycardia, hypertension, fever, diaphoresis, and tremor.

delirium tremens—the classic delirium associated with alcohol withdrawal.

delusion—a fixed false belief; common in psychiatric illness, and also a feature of neurologic disorders affecting the temporal lobes.

dementia—an acquired and persistent impairment in intellectual ability that affects at least three of the following domains: memory, language, visuospatial perception, praxis, calculations, conceptual or semantic knowledge, executive function, personality or social behavior, and emotional awareness or expression.

denial—explicit refusal to acknowledge disability, usually seen with right hemisphere lesions (see *anosognosia* and *anosodiaphoria*).

developmental dyslexia—difficulty learning to read despite adequate intelligence, motivation, and educational opportunity.

diffuse axonal injury—widespread shearing injury of white matter (and its axons) in the brainstem, hemispheres, and corpus callosum, common to all forms of traumatic brain injury; also called traumatic axonal injury.

disconnection syndrome—a neurobehavioral syndrome due to interruption of cerebral white or gray matter structures linking cortical areas; callosal and intrahemispheric disconnection syndromes are recognized.

disinhibition—diminished control over inappropriate limbic impulses, typically seen with bilateral orbitofrontal lesions.

disorientation—inability to recall one's name, the place, and the date; the latter two errors are commonly used as indicators of recent memory dysfunction.

distractibility—susceptibility to distraction by inessential elements of experience; often seen with frontal lobe lesions.

distributed neural networks—large-scale ensembles of brain neurons devoted to distinct cognitive or emotional functions that include integrated gray and white matter regions within or between the cerebral hemispheres (see *frontal-subcortical circuits*).

dressing apraxia—difficulty with dressing due to misunderstanding of the relationship of clothing with the shape of the body; usually seen with right parietal lesions.

dysarthria—an acquired disorder of speech, representing damage to the motor system subserving articulation.

dysphonia—an acquired disorder of voice, reflecting damage to the larynx.

echolalia—involuntary repetition of words or phrases spoken by the examiner; typically seen in mixed transcortical aphasia but also in Alzheimer's disease, schizophrenia, and other disorders.

echopraxia—the imitation of gestures without the direction to do so; often seen with frontal lobe lesions.

epilepsy—a disorder characterized by recurrent unprovoked seizures due to an abnormal electrical excitability of the cerebral cortex; a synonym is seizure disorder.

erotomania—a delusion typically seen in women who believe that an older, influential man is in love with them despite all evidence to the contrary; also called de Clérambault's syndrome.

executive dysfunction—difficulty with planning, sustaining, monitoring, and completing complex behaviors, typically seen with bilateral dorsolateral frontal damage.

extinction—failure to report a stimulus on the side of the body opposite to a cerebral hemispheric lesion when simultaneous stimuli are presented to both sides; extinction can be elicited in the visual, auditory, and tactile systems.

forced normalization—a term indicating the inverse relationship between seizure control and psychosis in temporal lobe epilepsy; also known as alternative psychosis.

formication—a tactile hallucination characterized by the sensation of insects crawling on the skin.

Fregoli syndrome—a delusional misidentification in which a patient believes a persecutor is taking on the appearance of a stranger; seen with bifrontal and right hemisphere lesions.

frontal alexia—a syndrome of literal alexia (in which letter reading is better than word reading) associated with left inferior frontal lesions; also known as anterior alexia or the third alexia.

frontal-subcortical circuits—a group of parallel circuits comprising integrated regions of the frontal cortex, basal ganglia, thalamus, and their white matter connections that are devoted to comportment, motivation, and executive function (see *distributed neural networks*).

frontotemporal lobar degeneration—a cortical dementia involving degeneration of the frontal and temporal lobes; its variants are frontotemporal dementia, progressive nonfluent aphasia, and semantic dementia.

gait apraxia—a disorder of gait not due to primary motor or sensory disturbance; often seen with bilateral frontal lobe dysfunction related to hydrocephalus.

Gerstmann's syndrome—acalculia, agraphia, finger agnosia, and right-left disorientation; classically associated with left inferior parietal lesions (see *angular gyrus syndrome*).

Geschwind syndrome—the interictal personality syndrome of some patients with temporal lobe epilepsy.

global aphasia—nonfluent speech with poor auditory comprehension, repetition, and naming, typically due to large left perisylvian lesions.

hallucination—a sensory experience without an external stimulus; seen in a wide variety of neurologic as well as psychiatric disorders.

hemineglect—neglect (see below) for sensory stimuli contralateral to the side of a cerebral lesion; more common with right than left hemisphere lesions.

hypergraphia—a tendency to produce excessive amounts of written material in the form of journals, diaries, poetry, and the like; seen in some patients with temporal lobe epilepsy.

hypophonia—an abnormally low volume of speech, seen in Parkinson's disease and other subcortical gray matter diseases.

ictal psychosis—psychosis occurring as a result of continuous seizure activity, most often complex partial status epilepticus of temporal lobe origin.

ideational apraxia—difficulty executing a series of learned movements even though the individual acts can be performed in isolation; seen with left parietal or diffuse cerebral involvement.

ideomotor apraxia—difficulty executing a learned movement to verbal command; seen with left perisylvian lesions.

illusion—a misperception of a stimulus; seen in clinical settings similar to those that predispose to hallucinations.

inattention—deficient ability to respond to salient external stimuli while excluding others; inattention may be bilateral, as in confusional states, or unilateral, as in focal (usually right hemisphere) lesions.

jargon aphasia—severe Wernicke's aphasia with rapid paraphasic speech and abundant neologisms.

Klüver-Bucy syndrome—the combination of hypersexuality, placidity, hyperorality, "psychic blindness," and hypermetamorphosis; seen after bilateral anterior temporal lobe lesions in higher primates and humans.

Korsakoff's amnesia—chronic amnesia that follows many cases of untreated Wernicke's encephalopathy; a more common but less accurate term for this syndrome is Korsakoff's psychosis.

limb kinetic apraxia—impairment of delicate motor acts in a limb contralateral to a premotor lesion; also called melokinetic or innervatory apraxia.

locked-in syndrome—a state of quadriplegia with spared vertical eye movements, eye blinking, and cognition, seen with large pontine lesions; a synonym is the de-efferented state.

macropsia—a visual illusion in which the image appears to be increased in size; most often seen with temporal lobe seizures and migraine.

metamorphopsia—a visual illusion in which the image appears to be distorted in shape; most often seen with temporal lobe seizures and migraine.

micrographia—a tendency to produce very small handwriting, associated with Parkinson's disease and other forms of parkinsonism.

micropsia—a visual illusion in which the image appears to be decreased in size; most often seen with temporal lobe seizures and migraine.

mild cognitive impairment—a memory disorder characterized by a memory complaint, normal activities of daily living, normal general cognitive function, abnormal memory for age, and no dementia; often, but not always, a precursor to Alzheimer's disease.

minimally conscious state—a condition of severely impaired level of consciousness in which minimal but definite behavioral evidence of self- or environmental awareness is demonstrable (see *persistent vegetative state*)

moria—an inappropriately excited and childish affect seen in patients with bilateral orbitofrontal lesions.

motor impersistence—inability to persist at an intended motor activity; seen with right cerebral lesions, particularly in the right frontal lobe.

mutism—failure to speak, seen in severe aphasia, acute aphemia, advanced dementia, akinetic mutism, and a variety of psychiatric disorders.

neglect—failure to notice, report, or respond to stimuli presented opposite to the side of a cerebral lesion.

neologism—a type of paraphasic error in which a new and meaningless word is produced; seen in aphasia, particularly fluent aphasia, and schizophrenia.

nonaphasic misnaming—a syndrome in patients with acute confusional state characterized by selective naming errors related to a personalized reference system.

nonfluency—a feature of spontaneous speech in aphasia characterized by reduced phrase length (five words or less), agrammatism, impaired linguistic prosody, and dysarthria; typical of Broca's aphasia and other nonfluent aphasias.

object agnosia—a form of visual agnosia characterized by the inability to recognize familiar objects; seen with left or bilateral occipitotemporal lesions.

ocular apraxia—inability to direct voluntary gaze to a target visual stimulus, also called oculomotor apraxia or psychic paralysis of gaze; a component of Balint's syndrome.

optic aphasia—inability to name visually presented objects but with preserved recognition, associated with left posterior subcortical lesions; a mild form of associative visual agnosia.

optic ataxia—impairment of limb movements under visual guidance; a component of Balint's syndrome.

Othello syndrome—a delusion in which a person harbors the unjustified conviction of a spouse's infidelity; also called delusional or pathological jealousy.

palilalia—involuntary repetition words, phrases, or syllables spoken by the patient; often seen in Parkinson's disease (see *echolalia*).

palinacousis—persistence or recurrence of an auditory experience after the auditory stimulus is no longer present; usually seen with temporal lobe lesions.

palinopsia—persistence or recurrence of a visual image after the visual stimulus is no longer present; usually seen with right occipitoparietal lesions.

paraphasia—an abnormality of aphasic speech characterized by letter or word substitutions; a paraphasic error can be literal (phonemic), verbal (semantic), or neologistic.

peduncular hallucinosis—formed visual hallucinations of a benign or entertaining nature associated with lesions of the cerebral peduncle or nearby midbrain structures.

perisylvian aphasias—the aphasias resulting from damage around the Sylvian fissure and characterized by impaired repetition: Broca's aphasia, Wernicke's aphasia, conduction aphasia, and global aphasia.

perseveration—continuation or recurrence of an activity without an appropriate stimulus; frequently seen with dorsolateral frontal lesions but can also occur with lesions in other regions.

persistent vegetative state—a condition of preserved arousal but absent cognition and emotion related to extensive damage of the cerebral hemispheres (see *minimally conscious state*).

postconcussion syndrome—headache, inattention, memory disturbance, fatigue, dizziness, insomnia, anxiety, and depression after concussive traumatic brain injury.

postictal psychosis—psychosis occurring in the period following a complex partial seizure of temporal lobe origin, or a cluster of these seizures.

post-traumatic amnesia—an impairment of new learning following the recovery of consciousness after traumatic brain injury.

primary progressive aphasia—a term used to refer to either progressive nonfluent aphasia or semantic dementia, linguistic forms of frontotemporal lobar degeneration.

progressive nonfluent aphasia—the variant of frontotemporal lobar degeneration featuring anomia and nonfluent aphasia resulting from involvement of the left frontal lobe.

prosopagnosia—a form of visual agnosia characterized by inability to recognize familiar faces; seen with right-side or bilateral occipitotemporal lesions.

pseudobulbar affect—involuntary laughing or weeping in patients with disinhibition of facial musculature due to bilateral corticobulbar damage; also called pathological laughter and crying, or emotional incontinence.

psychomotor retardation—slowing of mental and motor activity, commonly seen in depression; also used to describe apathy and abulia with bilateral medial frontal lesions and bradyphrenia with basal ganglia diseases.

reduplicative paramnesia—a delusional belief that a familiar place has been relocated to another site; seen with bifrontal and right hemisphere lesions.

retrograde amnesia—inability to remember information acquired shortly before the onset of an amnesic disorder.

schizophreniform psychosis—a psychotic disorder developing in individuals that resembles schizophrenia but is putatively related to preexistent temporal lobe epilepsy.

seizure—an abrupt, involuntary neurologic event caused by a paroxysmal and excessive electrical discharge of cerebral cortical neurons.

semantic dementia—the variant of frontotemporal lobar degeneration featuring fluent aphasia and impaired word-meaning knowledge resulting from left temporal lobe degeneration.

simultanagnosia—inability to recognize all elements of a visual scene simultaneously; a component of Balint's syndrome.

stimulus-bound behavior—inappropriate activity based on excessive attention to insignificant or irrelevant stimuli; often encountered with dorsolateral frontal lesions.

strategic infarct dementia—dementia following a single infarct in one cerebral area that is particularly important in the representation of cognitive functions, such as the left angular gyrus or the thalamus.

stroke—an acute interruption of flow within a cerebral blood vessel causing focal destruction of brain tissue; a less preferred synonym is cerebrovascular accident.

stupor—a disorder of arousal characterized by unresponsiveness that can only be overcome by vigorous and repeated stimuli.

temporal lobe epilepsy—a type of seizure disorder in which the irritable cerebral focus lies in the temporal lobe; closely related to complex partial seizure disorder and the older concept of psychomotor seizure disorder.

transcortical aphasias—the aphasias resulting from damage outside the perisylvian region and characterized by preserved repetition: transcortical motor aphasia, transcortical sensory aphasia, anomic aphasia, and mixed transcortical aphasia.

transient global amnesia—a transient, reversible amnesia, most likely related to ischemia of the medial temporal lobes.

verbal apraxia—a term used to describe impaired speech that falls intermediate between dysarthria and aphasia.

viscosity—a tendency in some patients with temporal lobe epilepsy to be verbose, repetitive, and detailed; also known as stickiness.

visuospatial dysfunction—acquired difficulty with the interpretation and representation of visuospatial information, often referred to as constructional apraxia.

Wernicke's aphasia—fluent language with poor auditory comprehension typically due to a lesion in the posterior portion of the left superior temporal gyrus.

Wernicke's encephalopathy—a disorder seen most often in alcoholics with thiamine deficiency, and consisting of acute confusional state, ophthalmoplegia, and gait ataxia.

Wernicke-Korsakoff syndrome—the combination of Wernicke's encephalopathy and Korsakoff's amnesia.

white matter dementia—a dementia syndrome associated with diffuse or multifocal cerebral white matter damage from disease, injury, or intoxication.

witzelsucht—inappropriate jocularity seen in patients with bilateral orbitofrontal lesions.

INDEX

Illustrations are indicated by page numbers in italics.